75
FSG

ALSO BY NATHANIEL RICH

Losing Earth

King Zeno

Odds Against Tomorrow

The Mayor's Tongue

SECOND NATURE

SECOND
NATURE

SCENES FROM A WORLD REMADE

NATHANIEL RICH

MCD

Farrar, Straus and Giroux
New York

MCD
Farrar, Straus and Giroux
120 Broadway, New York 10271

Some of this material previously appeared, in substantially different form, in the following publications: *The New York Times Magazine* ("Dark Waters," "Here Come the Warm Jets," "Frankenstein in the Lower Ninth," "Pigeon Apocalypse," Parts I and III of "Bayou Bonjour," and "The Immortal Jellyfish"); *The New Republic* (Part II of "Bayou Bonjour"); *Vice* ("The Wasting"); and *Men's Journal* ("Aspen Saves the World").

Grateful acknowledgment is made for permission to reprint the following previously published material:
Lyrics from "Scarlet Medusa Chorus" and "I Am Shin Kubota," by Shin Kubota, used with permission.

Library of Congress Cataloging-in-Publication Data
Names: Rich, Nathaniel, 1980– author.
Title: Second nature : scenes from a world remade / Nathaniel Rich.
Description: First edition. | New York : MCD / Farrar, Straus and Giroux, 2021.
Identifiers: LCCN 2020049011 | ISBN 9780374106034 (hardcover)
Subjects: LCSH: Nature—Effect of human beings on. | Human beings—Effect of environment on. | Environmentalism.
Classification: LCC GF75 .R52 2021 | DDC 304.2/8—dc23
LC record available at https://lccn.loc.gov/2020049011

Our books may be purchased in bulk for promotional, educational, or business use. Please contact your local bookseller or the Macmillan Corporate and Premium Sales Department at 1-800-221-7945, extension 5442, or by email at MacmillanSpecialMarkets@macmillan.com.

www.mcdbooks.com • www.fsgbooks.com
Follow us on Twitter, Facebook, and Instagram at @mcdbooks

1 3 5 7 9 10 8 6 4 2

This is for Julian

CONTENTS

SECOND NATURE

Introduction: STRANGE VICTORY

Fort Bragg's Glass Beach is the most popular attraction on the Northern California coast. It receives more visitors than the Lost Coast, through which steep trails navigate cloud forests, waterfalls, and ocean panoramas. It gets far more traffic than the Mendocino Coast Botanical Gardens and Mendocino Headlands State Park. From the parking lot off Glass Beach Drive, tourists descend a steep staircase between graywacke cliffs to photograph a narrow cove that sparkles with turquoise and brown and ruby shards, buffed and rounded by the surf. Posted signs beg visitors—a couple of thousand a day during the summer—not to pocket the glass but they can't help themselves.

In 2012, J. H. "Cass" Forrington, a retired sea captain and the owner of the nearby International Sea Glass Museum, which displays more than three thousand poached pieces, led a campaign to "replenish" the beach with tons of broken glass. Forrington's argument rested on an ecological claim. Because the sea glass, which created habitat for microscopic marine life, had integrated into the local ecosystem, it deserved the same protections granted to the coast redwood, the mountain beaver, the red-legged frog.

The California Department of Fish and Wildlife is responsible for protecting and maintaining "natural communities for their intrinsic and ecological value and their benefits to people." The fate of Glass Beach hung on the definition of "natural." Forrington argued that California was legally bound to dump more glass on the sand. "To say the glass is not 'natural' is simply wrong," he wrote, in a manifesto littered with an unassailable profusion of quotation marks. "Because of the damage we can do to an overall habitat, we tend to think of ourselves as being somehow 'un-natural,' and 'outside' of 'nature,' but we are an integral part of 'nature' and we can also do great good."

The great good to which Forrington referred dates to 1949, when the beach was designated for use as a landfill. The tons of glass pebbles and ellipsoids that littered the cove were the remnants of beer bottles, taillight lenses, and Tupperware. For the next two decades the beach was known to locals as the Dumps. The only way to regain the beach's natural beauty, wrote Captain Forrington, was to bury it every year under a few more tons of trash.

In the end the Department of Fish and Wildlife was unpersuaded by Captain Forrington's definition of "nature"; it declined to intervene. Forrington would not be so easily defeated, however. In defiance, he continued to sell plastic bags of pre-tumbled glass to tourists who lugged them down the wooden stairs and emptied them onto the sand. Captain Forrington believed he was doing his part to save nature, or at least "nature."

Long after the last copy of the King James Bible has disintegrated and the Venus de Milo has gone to powder, the glory of our civilization will survive in misshapen, neon-flecked rocks called plastiglomerate: compounds of sand, shells, and molten plastic, forged when candy bar wrappers and bottle caps burn in campfires. Additional clues will be found in the ubiquity of cesium-137, the synthetic isotope produced by nuclear detonations; a several-thousand-year diminution in calcium carbonate deposition, the consequence of ocean acidification; and glacial ice cores (should glaciers remain) registering a dramatic spike of atmospheric carbon dioxide. Future anthropologists might not be able to learn everything there is to know about our culture from these geological markers but it will be a good start.

In the beginning, human beings tended to view nature as a mortal enemy—with wariness, dread, and aggression. The war began before we had even bothered to name our enemy. Already in the earliest literature, the assault is well under way, the bellicosity raw, the motives unquestioned. In "The Lord to the Living One's Mountain," Gilgamesh, terrified of death, decides he must perform a heroic feat to achieve immortality. As he can imagine nothing more honorable than the destruction of a virgin forest, Gilgamesh travels to the sacred Mountain of Cedar, beheads the demigod who defends it, and razes the forest to stubble, reserving the grandest tree for use as a gate to his city.

About seventeen centuries later, in Plato's *Phaedrus*, Socrates, reluctant to venture outside Athens' city walls, declares,

"I am a lover of learning, and the outdoors and trees have never taught me anything, whereas in the city there are people and they do teach me." Aristotle is more direct in *Politics*: "Nature has made all things specifically for the sake of man." In the Old Testament, "the wilderness" is a godless domain, the anti-Eden. As in: "He led you through the vast and dreadful wilderness, that thirsty and waterless land, with its venomous snakes and scorpions."

"Wilderness": from the Old English -ness + wild + deor, "the place of wild beasts." Samuel Johnson defined it as "a tract of solitude and savageness." William Bradford, a founder of Plymouth Colony, reacted to the New World with horror, calling it "hideous & desolate . . . full of wild beasts & wild men." The most widely collected work of the Enlightenment, Comte de Buffon's thirty-six-volume *Natural History*, proliferates with words like "grotesque," "nauseous," "pestilential," "terrible," and "filth."

Nature invited subjugation—for its own good. The American jurist James Kent extended this conceit to the human beings who had lived for millennia in harmony with the land as he sought to construct a legal basis for seizing territory from Native Americans. The continent, Kent argued, was "fitted and intended by Providence to be subdued and cultivated, and to become the residence of civilized nations." The gospel of Nature was a license to dominate, brutalize, and pillage—and feel proud of it.

Some of these examples come from Roderick Nash's totemic history, *Wilderness and the American Mind*, which describes how finally, in the nineteenth century, the terms of humanity's relationship with nature flipped. Scientists and philosophers began to question the premise that nature was a threat to civilization. They'd had it backward: civilization was a threat to nature. It had become obvious that humanity was winning its thousands-year war against nature in a rout. It was a costly victory, however. The prize was civilizational collapse.

This understanding was first articulated by Alexander von Humboldt, who was born in 1769, during the era in which human beings stopped fearing nature and took pride in their ability to master it. It was the age of the steam engine, the smallpox vaccine, the lightning rod. Timekeeping and measuring systems became standardized; the blank spaces remaining on world maps were shaded in. Even before Humboldt began his global tour, analyzing everything from wind patterns and cloud structures to insect behavior and soil composition, he intuited that Earth was "one great living organism where everything was connected." It is commonplace today to speak of "the web of life," but the concept was Humboldt's invention. It followed that the fate of one species might have cascading effects on others. Humboldt was among the first to warn of the perils of irrigation, cash crop agriculture, and deforestation. By 1800 he had come to realize that the damage wreaked by industrial civilization was already "incalculable."

Humboldt's insights were developed by acolytes like George Perkins Marsh (who warned that "climatic excess" might lead to human extinction); Charles Darwin (who plagiarized Humboldt in the final, crowning paragraph of *On the Origin of Species*); Ralph Waldo Emerson ("the whole of nature is a metaphor of the human mind"); and the besotted John Muir ("This sudden plash into pure wildness—baptism in Nature's warm heart—how utterly happy it made us!"). By the turn of the twentieth century, Americans increasingly began to see wilderness as a spiritual refuge from the mechanization of modern life. Horror had turned to infatuation.

Yet the romantic view of nature proved counterproductive. It encouraged the protection of natural cathedrals like Yosemite and Yellowstone while devaluing the pedestrian swaths of forest, swamp, and grassland that make up most of the country. Before long the cathedrals were besieged too, victims of political pragmatism. Theodore Roosevelt and Gifford Pinchot, first chief of the U.S. Forest Service, embraced a utilitarian approach to ensure that wilderness sanctuaries could be enjoyed by both hikers and oil prospectors. When such interests came into conflict, however, conservationists lost—most flagrantly in the battle over Yosemite's Hetch Hetchy Valley, dammed in 1923 to provide water to San Francisco.

"Engineering is clearly the dominant idea of the industrial age," wrote Aldo Leopold, the father of wildlife ecology, in 1938. "Ecology is perhaps one of the contenders for a new order . . . Our problem boils down to increasing the overlap of awareness between the two." Ecology, though the severe underdog, made tentative advances over the course of the twentieth century. By the first Earth Day, in 1970, it had birthed a new political movement. In the following decade, the politics of nature evolved to reflect a broader understanding of the interconnectedness of ecological threats. Concerns over air and water pollution, climate change, land development, resource extraction, species extinction, drought, wildfires, and roadside littering were consolidated under the rubric of "the environment." The definition has since expanded further to reflect the insight that ecological degradation, by exacerbating the inequalities that poison our society, degrades democracy itself. This realization has sounded the death rattle of the romantic idea that nature is innocent of human influence. We're innocent no longer.

What we still, in a flourish of misplaced nostalgia, call "the natural world" is gone, if ever it existed. Almost no rock, leaf, or cubic foot of air on Earth has escaped our clumsy signature. As Diane Ackerman has written, "It's as if aliens appeared with megamallets and laser chisels and started resculpting every continent. We've turned the landscape into another form of architecture; we've made the planet our sandbox."

No one has better articulated the incoherence of the nature ideal than the historian William Cronon in his transformational "The Trouble with Wilderness; or, Getting Back to the Wrong Nature." Cronon takes, more or less, Captain Forrington's position. Nature, he writes, "is quite profoundly a human creation . . . As we gaze into the mirror it holds up for us, we too easily imagine that what we behold is Nature when in fact we see the reflection of our own unexamined longings and desires." The idealization of wilderness is not merely a myth; it is antagonistic to the aims of any environmentalist. For if, in the future, something resembling wilderness is to survive, it will be only "by the most vigilant and self-conscious management."

Our most prized wilderness areas are already the beneficiaries of governmental regulation, political compromise, and the constant round of interventions euphemized as "land management." Even the rewilding movement, which preaches benign neglect to allow nature to recover at its own pace, acknowledges the need to meddle. *Wilding*, Isabella Tree's account of the transformation of her English estate into a nature refuge, details the installation of barbed wire, the importation of longhorn cattle and trapped deer, and generous applications of glyphosate. The most ambitious rewilding project, the biologist Edward O. Wilson's proposal, set forth in *Half-Earth*, to designate one-half of the planet a nature preserve, is based on the proposition that we have become "the architects and rulers of the Anthropocene epoch"—an echo of Descartes' "the lords and possessors of nature"—and must take responsibility for it. The creation of a Half-Earth would, after all, require political treaties, taxes, and armies.

We have followed Aldo Leopold's instruction to preserve "some tag-ends of wilderness, as museum pieces, for the edification of those who may one day wish to see, feel, or study the origins of their cultural inheritance." We've succeeded— calamitously. We have the tag ends and little else. One of the fundamental lessons of ecology is that isolated patches of wilderness starve to death.

The engineer and the ecologist have been enemies from the cradle. Since its founding as a discipline in the eighteenth century, civil engineering has sought to bring an unruly planet to heel—flattening infelicities of grade and angle, simplifying rugged terrain into a planar grid, routinizing chaos. But in recent decades a shift has begun. Engineers have designed buildings shaped like mountains to reduce their emissions, wind turbines that mimic whale fins to increase efficiency, bricks of bacteria that inhale carbon dioxide. They have achieved a more powerful control of nature through the imitation of nature.

Ecologists, meanwhile, have accepted that a threatened ecosystem requires steady interventive care, as might any patient in critical condition.

Two dovetailing observations by the novelist William Gibson describe the next chapter of this history. The first has hardened into platitude: "The future is already here—it's just not evenly distributed." The other is "soul delay," the idea that during long-distance flights the human body travels faster than the spirit: "Souls can't move that quickly, and are left behind, and must be awaited, upon arrival, like lost luggage." The uneasy sensation of waiting for your soul to catch up is what we call jet lag.

We now inhabit a similar lag: a nature lag. The future is already here, unevenly distributed. We recognize its hallmarks: rising sea levels, regular visitations of apocalyptic natural disasters, the forced migration of tens of millions, accelerating extinctions, coral bleaching, global pandemic. Also: cultured meat, reengineered coastlines, the reanimation of extinct species, bunny rabbits that glow fluorescent green. Our souls haven't caught up.

Even in the most optimistic future available, we will profoundly reconfigure our fauna, flora, and genome. The results will be uncanny. It will be difficult to remember that they will be no more uncanny than our carpeting of the American Southwest with lush lawns transplanted from the shores of the Mediterranean, our breast-augmented chickens, our taming of the world's most violent rivers. If our inventions seem eerie, it is only because we see in them a reflection of our desires. It is impossible to protect all that we mean by "natural" against the ravages of climate change, pollution, and psychopathic corporate greed, unless we understand that the nature we fear losing is our own.

The conservation of nature means the conservation of our identity: the parts of us that are beautiful and free and sacred, those that we want to carry with us into the future. If we don't defend those soft parts, all we'll have left are holograms of our worst instincts, automatons impersonating our nightmares, and a slow drift into a desert of biblical dimensions: *a tract of solitude and savageness.*

What follows are stories of people who ask difficult questions about what it means to live in an era of terrible responsibility. In the first part, "Crime Scene," a series of amateur detectives investigate crimes against nature. Confronted with the worst of humanity, they ask, who let it come to this?

The stories in "Season of Disbelief" are about people whose fundamental understanding of the physical world is mocked by a new reality. When our land, food, and climate no longer resemble anything we've known, how do we avoid losing our humanity too?

"We are as gods and might as well get good at it," wrote Stewart Brand in the *Whole Earth Catalog*. He has since revised this to "We are as gods and HAVE to get good at it." We know what it looks like to be *bad* at it. Margaret Atwood's

MaddAddam trilogy, the films of Alex Garland, Edward Bur-
tynsky's panoramas of industrial wastelands, the petri-dish
art of Suzanne Anker, and the biographies of monomaniacal
billionaires in Brunello Cucinelli T-shirts give some flavor of
it. The environmentalist's anxiety over "technofixes" has less
to do with the technology itself than with who exploits it.
"Technology is neutral," writes Roderick Nash. "The issue is
how it is used." Because we are not gods but primates plagued
by fear and hubris, impersonating divinities usually ends in
humiliation. In "As Gods," artists and engineers navigate unintended consequences, ethical cul-de-sacs, and their own
vanities as they struggle to create a more human future.

The trajectory of our era—this age of soul delay—runs
from naivete to shock to horror to anger to resolve. There is
no better avatar of this transformation than Robert Bilott, a
corporate defense lawyer who started as a man of DuPont's
America and became a man of the future.

Part I

CRIME SCENE

1

DARK WATERS

A few months before Robert Bilott made partner at Taft Stettinius & Hollister, he received a call from a cattle farmer in Parkersburg, West Virginia. Wilbur Tennant said that his cows were dying left and right. He was certain that the DuPont chemical company, which operated a site in Parkersburg more than thirty-five times the size of the Pentagon, was to blame. Tennant complained that he had tried to seek redress locally but DuPont about owned the entire town. He had been ignored not only by Parkersburg's lawyers but also by its politicians, journalists, and doctors. Bilott struggled to make sense of this. Tennant wasn't easy to understand: he spoke in a singing mountain dialect and was spitting mad besides. Bilott couldn't imagine how the farmer had gotten his phone number and he might have hung up had Tennant not blurted out the name of Bilott's grandmother.

Alma Holland White had lived in Vienna, a northern suburb of Parkersburg, where Bilott had visited her during his boyhood summers. In 1973 she brought him one weekend to a farm belonging to her friends the Grahams, who were neighbors to the Tennants. Bilott rode horses, milked cows, and watched Secretariat win the Triple Crown on TV. Of an itinerant, unpredictable childhood, this was one of Bilott's happiest memories. He had been seven years old.

When the Grahams heard in 1998 that Wilbur Tennant was looking for an environmental lawyer, they remembered that their friend's grandson had grown up to become one. They did not understand that Bilott was the wrong kind of environmental lawyer. He did not represent plaintiffs or private citizens. Like the other two hundred lawyers at Taft, a firm founded in 1885 with close ties to the family of President William Howard Taft, Ohio's leading Republican dynasty for more than a century, Bilott was a defense lawyer. He specialized in defending chemical companies. DuPont's lawyers were his colleagues. He respected the DuPont culture. The company had the money, the expertise, and the pride to do things the right way, so the notion of it recklessly poisoning a poor farmer seemed not only unprecedented but irrational. Still he agreed to meet the farmer. He would tell his colleagues that he did so out of loyalty to his grandmother. But it was also out of loyalty to some distant part of himself.

A week later, Wilbur Tennant—burly, six feet tall, jeans, plaid flannel shirt, baseball cap—arrived at Taft's headquarters in downtown Cincinnati with his wife, Sandra. The Tennants hauled cardboard boxes crammed with videotapes, photographs, and documents into the firm's glassed-in reception area on the eighteenth floor. They were shown to a waiting room in which they sat on gray mid-century modern couches beneath an oil portrait of one of

Taft's founders. Bilott's supervisor, a partner named Thomas Terp, was curious enough to join the meeting himself. Tennant was not, after all, the typical Taft client. "Let's put it this way," Terp would say years later. "He didn't show up at our offices looking like a bank vice president."

Wilbur Tennant explained that he and his four siblings had run the cattle farm since their father abandoned them as children. They had only seven cows then, two hundred chickens, and a fifteen-hundred-dollar mortgage. To survive, they had to forage in the hills for roots and berries. Over decades they steadily acquired land and cattle, investing every dollar they made back into the farm, until two hundred cows roamed more than six hundred hilly acres. The property would have been even larger had his brother Jim and Jim's wife, Della, not sold sixty-six acres in the early 1980s to DuPont. The company wanted a landfill for waste from its plastics factory near Parkersburg, called Washington Works, where Jim worked as a laborer, digging ditches, pouring concrete, and cleaning debris. Executives showed up at the property in a limousine, offering a deal. The Tennants did not want to sell, but Jim had been in poor health for years, mysterious ailments his doctors could not diagnose, and they needed the money.

DuPont rechristened the plot Dry Run Landfill, named after the creek that ran through it. Dry Run Creek flowed to a pasture where the Tennants grazed their cows. Not long after the sale, the cattle began to act deranged. They had been like pets to the Tennants, even family members. At the sight of a Tennant they would amble over, nuzzle, let themselves be milked. No longer. Now, they drooled uncontrollably. They birthed stillborn calves. Their teeth turned black. Their pink eyes glowered murderously. When they saw the farmers, they charged. After Della and her daughters encountered a cow in a death agony, making "the awfullest bellow you ever heard, the blood just gushing out of its nose and mouth and rectum," she refused to walk the property without a loaded gun. Three-quarters of the herd had died.

It wasn't just the cows: there were legions of dead fish, frogs pet dogs and cats, and deer. The deer died in strange ways. They lay down in groups, like members of a suicide cult. The Tennants stopped eating the deer after Jim, while dressing a buck, found that its guts were fluorescent green.

At Taft a VCR was wheeled into a windowless conference room and Wilbur loaded one of his videocassettes. The footage, shot on a camcorder, was grainy and intercut with static. Images jumped and repeated. The sound accelerated and slowed. It had the pace and palette of a horror movie.

In the opening shot the camera panned across the creek, taking in the surrounding forest, the white ash trees shedding their leaves, and the shallow, creeping water, before holding on what appeared to be a snowbank at an elbow in the creek. The camera moved in, revealing a mound of soapy froth.

"I've taken two dead deer and two dead cattle off this ripple," said Wilbur in voice-over. "The blood run out of their noses and out their mouths. They're trying to cover this stuff up. But it's not *going* to be covered up, because I'm going to bring it out in the open for people to see."

The camera tracked a large pipe tilted into the creek, discharging green bubbles. "This is what they expect a man's cows to drink on his own property," said Wilbur. "It's about high time that someone in the state department of something-or-another got off their *can*."

The video cut to a skinny red cow standing in hay, its hair patchy and its back humped—kidney malfunction, Wilbur speculated. Another blast of static was followed by a close-up of a dead black calf collapsed in the snow. Its eye sparkled a brilliant chemical blue. "One hundred fifty-three of these animals I've lost on this farm," said Wilbur. "Every veterinarian that I've called in Parkersburg will not return my phone calls or don't want to get involved." He sighed. "Since they don't want to get involved, I'll have to dissect this thing myself. I'm going to start at this head."

The video cut out momentarily. It returned with a close-up on the calf's bisected head against the snow. There followed portraits of the calf's blackened teeth, its dissected liver, heart, stomachs, kidneys, and gallbladder. Wilbur pointed out unusual textures and discolorations. "I don't even like the looks of them," he says. "It don't look like anything I've been into before."

Tennant explained to Bilott that he kept organs in his freezer in the hope that they might one day be tested in a lab. He had found giant tumors, collapsed veins, green muscles. What he didn't preserve, he burned. At night when it rained, the cattle bones shone in the dark like glow sticks.

Bilott spoke with the Tennants for several hours, watching video and reviewing photographs. He saw cows with stringy tails, malformed hooves, giant lesions protruding from their hides, and red receded eyes; cows suffering constant diarrhea, slobbering white slime the consistency of toothpaste, staggering bowlegged as if drunk. Wilbur invariably zoomed in on their eyes. "This cow's done a lot of suffering," he would say, his voice pinched with horror, as a blinking eye expanded to the size of the screen.

Bilott didn't know what to say. *This is bad*, he thought. *There's something really bad going on here.*

$$\oplus$$

Bilott agreed immediately to take the Tennant case. It was, he felt, "the right thing to do." But that didn't mean that Bilott felt that his previous work for Taft, representing chemical companies, had been the wrong thing to do. He hadn't really thought about the ethics of the job one way or the other, if he were to be honest about it.

Bilott spoke cautiously, softly, with a lawyer's aversion to making unqualified statements. Stress played around the corners of his eyes. He was at great pains to conceal the furious energies behind his composed demeanor, but on occasion, when speaking about some injustice done to him or his clients, an inner anger flashed through a sudden wince or scowl. Soft-spoken, milk complected, with primly combed hair that grayed at the temples and a strict personal dress code of anonymous ties and cleanly pressed dark suits, Bilott convincingly played the role of interchangeable corporate lawyer. But it was a role—one for which he'd had to study and rehearse and hone. Unlike most of his Taft colleagues, he had not attended an Ivy League college or law school. He did not hold a membership to, or know the difference between, the Camargo and the Kenwood country clubs. His father had been a lieutenant colonel in the air force, and Bilott spent most of his childhood moving among bases near Albany, New York; Flint, Michigan; Newport Beach, California; and Wiesbaden, West Germany. He had attended eight schools before graduating from Fairborn High, near Ohio's Wright-

Patterson Air Force Base. As a junior, he received a recruit-
ment letter from a tiny liberal arts school in Sarasota called
the New College of Florida, which graded pass-fail and allowed students to design their own curricula. That sounded
good to him. The friends he made in Sarasota were idealis-
tic, progressive—ideological misfits in Reagan's America. He
met individually with his professors, who emphasized the
value of critical thinking. He learned to question everything
he read, to refuse to take anything at face value, to ignore
the opinions of others. That philosophy confirmed his fundamental sense of the world and gave him the language to
articulate it. Bilott studied political science and wrote his
thesis about the rise and fall of Dayton. He hoped to become
a city manager.

But his father, who late in life enrolled in law school,
encouraged Bilott to do the same. To the surprise of his professors, Bilott withdrew from a doctoral program in public
administration to attend law school at Ohio State. His favorite course was environmental law. It was the rare legal field in
which he felt he "could make a difference." When he accepted
an offer from Taft, his mentors and friends from New College
were aghast. They called him a sellout. Bilott didn't see it that
way. He had just wanted to get the best job he could. He had
never known anyone who had worked in corporate practice,
but his father told him that the larger and wealthier a firm
was, the more opportunities he'd get. That made sense to
Bilott, though he didn't really give it much thought one way
or another.

At Taft, he volunteered to join Thomas Terp's environmental team. It was, as he had suspected, a moment of tremendous opportunity for environmental lawyers. Ten years

earlier, Congress had passed the legislation known as Super-fund, which financed the emergency cleanup of hazardous-waste dumps. Superfund created an entire subfield within environmental law, one that required a deep understand-ing of the new regulations to guide negotiations between the government and private companies. This was a lucrative line of business for Taft, if not a particularly alluring one to recruits—with the exception of Rob Bilott.

As an associate, Bilott was asked to determine which companies contributed which toxins and hazardous wastes in what quantities to which sites. He took depositions from factory workers, searched public records, and organized vast quantities of historical data. He became an expert on the Environmental Protection Agency's regulatory framework, the Safe Drinking Water Act, the Clean Air Act, the Toxic Substances Control Act. He mastered the chemistry of the pollutants, even though chemistry had been his worst subject in high school. He learned how the companies managed haz-ardous waste, how the laws were applied, and how to protect his clients. He became the consummate insider.

Bilott was proud of his work. Most of his clients, he be-lieved, wanted to do the right thing. It was his job to help show them how to do it. He worked long hours, and some of the other associates began to worry about him. It was obvi-ous that he knew few people in Cincinnati and he made no time to meet anyone. A colleague on Terp's environmental team took it upon herself to introduce him to a childhood friend named Sarah Barlage. She was a lawyer, too, at another downtown firm, where she defended corporations against workers' compensation claims. Bilott joined the friends for lunch at Arnold's Bar and Grill, the oldest tavern in downtown Cincinnati. Years later, Sarah would have no memory of Bilott opening his mouth. Her first impression was that he was "different, not like the other lawyers." She didn't mind that he was quiet; she was chatty, so she figured they complemented each other. Bilott didn't talk much on their first date either, because they went to see a movie, *Cape Fear.* Later they confessed to each other that they hated going to movies. They married in 1996.

The first of their three sons was born two years later. Bilott felt secure enough at Taft for Barlage to quit her job and raise their children full-time. Terp, his supervisor, recalled him then as "incredibly bright, energetic, tenacious and very, *very* thorough." He was a model Taft lawyer. Partnership was in hand.

Then Wilbur Tennant called.

The Tennant case put Taft in a highly unusual position. It was awkward, taking on DuPont, but Thomas Terp assumed the case would be easily resolved. He defended Bilott against anxious colleagues, explaining that the occasional plaintiff's suit made Taft a better defense firm, just as a photographer might sit for a portrait or a CEO might take a shift on the assembly line. It was the Taft way, he said, though Bilott's colleagues were not convinced.

To bring a lawsuit in West Virginia, Bilott needed a local attorney. He turned to Larry Winter, a personal injury lawyer who could not have been more different socially—garrulous, charming, loose—but shared Bilott's knowledge of how corporate giants operated. For many years, Winter was himself a

DuPont lawyer—a partner at Spilman Thomas & Battle, which represented DuPont in West Virginia. Winter was stunned that Bilott would sue DuPont while remaining at Taft— "inconceivable," he would say—but he was happy to join the case.

Bilott did for the Tennants what he would have done for any corporate client. He pulled permits, studied land deeds, and requested all documentation related to the Dry Run Landfill, including the chemicals DuPont dumped into it. Bilott and Winter formally filed a federal suit against Du-Pont in the summer of 1999 in the Southern District of West Virginia. It alleged that the company had violated its permits and contaminated the Tennants' property. Within a week, Bilott received a call from DuPont's in-house lawyer Bernard Reilly.

It seemed like a stroke of fortune: Bilott had known Reilly for years and admired him. Reilly spoke to him in the avun-cular manner of a senior colleague, a fellow club member. Reilly wanted to help the young lawyer with his grandma's cause; he really did. To that end, Reilly had good news: Du-Pont had already begun, at no small expense, its own study of the site, in partnership with the EPA. Six veterinarians— three chosen by DuPont, three by the EPA—would deter-mine the source of the cattle's problems. Bilott, reassured that a solution was in hand, agreed to wait for the results be-fore requesting any further documents. Reilly said it would take a couple of weeks.

It took six months. The veterinarians concluded that Du-Pont was not to blame for the dying cattle. Despite a rigorous testing of the creek, they had detected "no evidence of toxic-ity." The culprit, instead, was poor husbandry: "poor nutrition, inadequate veterinary care and lack of fly control." The Tennants didn't know how to raise cattle. If the cows were dying, it was their own fault.

This did not sit well with the Tennants, who had begun to suffer the consequences of antagonizing Parkersburg's main employer. They complained to Bilott that men in trucks parked across the road from their property and photographed them. They came home from church to find their personal files tossed around the room. Helicopters flew over their homes, hovering so low that picture frames fell off the walls. Lifelong friends began to ignore the Tennants on the street and walked out of restaurants when the Tennants entered. When confronted, they would tell the Tennants, "I'm not allowed to talk to you," it being understood that DuPont's word was, in Parkersburg, as binding as the Word. The Tennants changed congregations four times. Wilbur Tennant, his freezer packed with bovine organs, grew increasingly paranoid—"ready to kill," said Della. When the helicopters droned overhead, he stood beside his truck with his .25-06 rifle, screaming at the sky.

The Tennants called the office nearly every day but Bilott had little to tell them. He couldn't blame them for getting angry. He was angry too. The cattle report, he realized much too late, was nothing more than a delay tactic, and a successful one at that, costing him six months of work. He asked the judge to postpone the trial; he needed time to search for documents that might explain what was responsible for the plague at the Tennants' farm.

After the cattle report, DuPont handed off the case to Spilman, the West Virginia law firm where Larry Winter had been a partner. It seemed like another break; Winter remained

cordial with his former colleagues. They agreed without
hesitation to the requests for DuPont's internal records. Bilott made a point of examining every document himself—
correspondence with regulators, permit applications, land
deeds—but he could find nothing that contradicted the cattle
study. When he asked for even more information, files related
to the chemicals used at Washington Works, the Spilman
lawyers became less cordial. They stopped cooperating; when
they could delay, they did. But it was not until Bernard Reilly,
whom Bilott assumed had left the case months earlier, called
Bilott's boss to complain, telling him to "back off with all
this unnecessary discovery," that Bilott realized he was onto
something. In August 2000, more than sixty thousand documents into his treasure hunt, he found it.

He found it in a letter that DuPont had sent to the EPA less
than two months earlier. The author, a director of "Applied
Toxicology and Health," referenced a "blood serum" study of
employees at Washington Works. The subject of the study
was a chemical with a cryptic name: ammonium perfluo-
rooctanoate, or APFO. In all his years working with chemical companies, Bilott had never heard of APFO. It did not
appear on any list of regulated materials, nor could he find it
in Taft's in-house library, with its chemical dictionaries and
exhaustive databases of hazardous substances. A chemistry
expert that Taft retained did, however, vaguely recall an article in a trade journal about a similar-sounding compound:
perfluorooctane sulfonate, a soap-like agent used by the technology conglomerate 3M in the fabrication of Scotchgard. Two

months earlier—right before the EPA letter—3M had announced it had stopped manufacturing it.

Bilott asked DuPont for every document in its possession related to the substance. DuPont refused. Bilott requested a court order to force them. Dozens of boxes of unorganized files began to arrive at Taft's headquarters: internal correspondence, medical reports, confidential studies. They came by the pallet and the luggage cart, more than one hundred thousand pages in all, some half a century old. This was a technique that lawyers called "burying them in paper" but for Bilott the boxes might have been wrapped in tinsel and ribbons. He spent the next few months on the floor of his office, poring over the documents and arranging them in chronological order. He gave up lunch breaks and stopped answering his phone. His secretary explained to callers that Mr. Bilott could not hope to reach his telephone in time, because he was trapped on all sides by boxes. Bilott began to get the feeling that he was the first person to have ever gone through all the files. It was clear, at the very least, that nobody at DuPont, or Spilman, had bothered to review what they were sending him. As he later put it, "I started seeing a story."

Bilott was given to understatement. ("To say that Rob Bilott is understated," said his colleague Edison Hill, "is an understatement.") The story that Bilott began to see, cross-legged on his office floor, would have been a tragedy had it offered any form of catharsis. It was astounding in its breadth, specificity, and sheer brazenness. Bilott would say he was shocked, but that was another understatement. He could not believe the scale of incriminating material that DuPont had gifted him. "It was one of those things where you can't believe you're reading what you're reading," he said. "That it's actually been put in writing. It was the kind of stuff you always heard about happening but you never thought you'd see written down."

⊕

On the flood wall at Point Park, on the bank of the Ohio a few miles upriver from Washington Works, is painted, in tall white letters,

Welcome to Parkersburg, W.V.
"Let's be friends."

Recently someone painted over the second line. "To deface [the flood wall] in such a manner is appalling to me," a resident wrote to the local editorial page. "I sincerely hope this was an act of vandalism that will soon be removed."

Those who live in Parkersburg and the neighboring towns have strange memories about the water. Sandra Follett remembers that when she was a teenager, she would sneak on summer nights with her friends onto DuPont property to visit a warm pond populated by two-headed frogs. The boys caught the little monsters and the girls screamed and the couples made out to the sound of the deformed frogs ribbiting out of both of their heads.

When Mike Smalley moved to the area with his wife, Linda, in the late 1980s, they heard rumors that DuPont polluted the drinking water, but they didn't believe it. Smalley cried at the memory. "Linda carried a big water jug. She'd fill it up when she got home from work and drink it all night. Me and the boys, we didn't drink water. I drank Coke; they drank Mountain Dew. But I bet she drank two gallons a day." Later, after she got sick, Linda would hang her head and say, "Look at all of the water I drank."

Darlene Kiger remembered the dying pets. Her mother kept buying parakeets, and they kept dying for no reason. Dogs died constantly. After her neighbor's dog developed gigantic tumors, she adopted a poodle that began sprouting tumors of her own. Kiger's own Maltese, whom she named Dog, did not get cancer. He went blind. Her children laughed when Dog ran into telephones, but they stopped laughing when Dog began to fall off chairs and convulse with seizures. They didn't adopt any dogs after that.

It wasn't only the animals. Darlene knew three men in their twenties diagnosed with testicular cancer. Every child in Parkersburg seemed to have asthma. She knew an old lady whose five-year-old granddaughter's teeth had turned black.

Bilott traced the story back to 1951, when DuPont started purchasing from 3M a chemical called perfluorooctanoic acid, or PFOA—the more common name for APFO. DuPont wanted PFOA for use in the manufacturing of Teflon. 3M had invented PFOA, which usually took the form of a preternaturally lubricious liquid, just four years earlier. It was used to keep coatings from clumping during production. Although PFOA was not classified as a hazardous substance, 3M recommended that it be incinerated or sent to chemical-waste facilities. DuPont's scientists drew up its own instructions for PFOA disposal. They specified it was not to be flushed into surface water or sewers and absolutely not, under any circumstances, into the public water supply.

Bilott learned that 3M and DuPont had been conducting secret medical studies on PFOA for more than four decades. In 1961, DuPont researchers found that exposure to the chemical could increase the size of the liver in rats and rabbits. It latched onto plasma proteins in the blood, hitchhiking the circulatory highway to tour each bodily organ. In the 1970s, DuPont discovered that the blood of Washington Works employees contained high concentrations of PFOA. Though federal law obliged companies to disclose evidence of any "substantial" human health risks caused by their products, DuPont did not report any of this to the EPA.

3M, which continued to serve as the supplier of PFOA to DuPont and other corporations, discovered between 1978 and 1981 that exposure to PFOA killed rabbits, beagles, and rhesus monkeys; when rats ingested the substance, their offspring had birth defects. After 3M shared this information, DuPont tested the children of pregnant employees in its Teflon division. Of seven births, two had eye defects. DuPont neglected to make this information public.

It was understood by then that PFOA's chemical structure made it uncannily resistant to degradation. It did not break down or metabolize. Like radiation exposure or atmospheric carbon, PFOA kept accumulating, in perpetuity. Every drop—every part per trillion—counted. Its durability, slipperiness, and immortal persistence, the qualities that made it magically useful in soaps and adhesives, made it nightmarishly toxic. You didn't need to drink PFOA by the gallon to be poisoned. Minimal quantities, sustained over time, would do the trick.

DuPont tried to determine whether PFOA could be detected in the air, land, and public water supply. For decades it had ignored its own guidelines and pumped hundreds of thousands of pounds of PFOA powder through the outfall pipes of the Parkersburg facility into the Ohio River. It dumped thousands of tons of PFOA-laced sludge into "digestion ponds": open, unlined pits on the Washington Works property, from which the chemical seeped into the ground. PFOA dust, the compound's most toxic incarnation, vented from factory chimneys.

In 1984, DuPont learned that PFOA dust was settling well beyond the property line and that PFOA sludge had entered the local water table. DuPont called a meeting at its headquarters in Wilmington, attended by nine senior executives and its lead toxicologist. DuPont's medical and legal teams proposed the immediate discontinuation of PFOA. The executives were unpersuaded; having produced PFOA for more than three decades, they were already legally responsible for damages; continued use would not create significant new liability. Besides, as one executive put it, elimination of PFOA would "put the long-term viability of the business segment on the line." In the following years DuPont boosted its production of PFOA with the wild abandon of a fugitive indulging in a final bender.

In 1991, DuPont scientists determined an internal safety limit for PFOA concentration in drinking water: one part per billion. The same year, DuPont found the local water contained PFOA levels three times as high. Again it declined to make the information public. Instead it decided to dispose of the toxic sludge that gurgled in open pits on company property. Fortunately they had just acquired sixty-six acres from a low-level employee that would do perfectly. In DuPont's files Bilott found property maps of the Tennant farm.

By the 1990s, Bilott discovered, DuPont understood that PFOA caused testicular, pancreatic, and liver cancer in lab animals. Exposure to PFOA was linked to DNA damage in one study, to prostate cancer in another. DuPont at last hastened to develop an alternative to PFOA. In 1993 an inter-office memo announced that "for the first time, we have a viable candidate" that appeared to be less toxic and did not remain in the body nearly as long. After another round of discussions at headquarters, DuPont again decided against a change. The risk was too high: products manufactured with PFOA were worth a billion dollars in annual *profit*. After 3M ended production of PFOA, DuPont built a new factory to make the substance itself.

DuPont produced more PFOA every year, with emissions peaking in 1999, the final year accounted for by the data. PFOA was used not only in the production of Teflon but in McDonald's French fry cartons, pizza take-out boxes, and microwave popcorn bags. PFOA and its chemical cousins were essential components of industrially produced furniture and carpets, camping gear, raincoats, aircraft fluids, X-ray film production, semiconductors, Post-its and other adhesive paper goods, and the firefighting foams used by the U.S. military. Seen from one vantage, fluorochemicals were the glue that held modern society together; from another, they were the slippery substances that made it dissolve.

In later document tranches sent by computer disk, Bilott discovered a series of messages that Bernard Reilly had written on his company BlackBerry. In notes to his son, Reilly complained about his bosses. "I can tell my clients 'I told you so,'" he wrote, "but that is small pleasure, pretty sad they are so clueless—guess they think folks like to drink our stuff." Elsewhere he lamented that DuPont "did not want to deal with this issue in the 1990s, and now it is in their face, and some are still clueless. Very poor leadership . . ." And he wrote, "The shit is about to hit the fan in WV. The lawyer for the farmer family finally realizes the surfactant issue . . . Fuck him."

The Tennant case suddenly seemed a minor aspect of the story, and one that could now be resolved. By 1990, Bilott learned, DuPont had dumped more than seven thousand tons of PFOA sludge into Dry Run. DuPont's scientists tested the creek and found it had sixteen hundred times the concentration of PFOA that they deemed safe. DuPont did not report this to the Tennants at the time, nor did it disclose the fact in the cattle report that it helped to produce a decade later—the report that blamed poor husbandry for the deaths of the Tennants' cows. Bilott had what he needed.

In August 2000, Bilott called Bernard Reilly and explained that he knew what was going on. It was a brief conversation. DuPont agreed to settle with the Tennants and pay Taft's contingency fee. The whole business might have ended right there. Bilott, however, was not satisfied. He was furious.

DuPont was nothing like the corporations he had represented at Taft. For decades, DuPont had known that PFOA was harmful, but it dumped gargantuan volumes of the stuff into the water and air and land anyway. Bilott had seen what PFOA had done to cattle. What was it doing to the tens of thousands of people who drank it daily from their taps? What did the insides of their heads look like? Did their internal organs glow fluorescent green?

Bilott spent months drafting a public brief, nearly a thousand pages long, against DuPont. His colleagues called it "Rob's Famous Letter." Citing "an imminent and substantial threat to health," it demanded immediate federal action to regulate PFOA and provide clean water to those living near the factory. On March 6, 2001, Bilott sent the letter to the director of every relevant regulatory authority, as well as the U.S. attorney general. DuPont reacted quickly, requesting a gag order to block Bilott from providing the government with the information he had uncovered in the Tennant case. A federal court denied the order. Bilott sent his entire case file to the EPA.

With the Famous Letter, Bilott crossed a line. Though nominally representing the Tennants—their settlement had yet to be finalized—Bilott spoke for the public, claiming extensive fraud and criminality. He had become a threat not merely to DuPont but also, in the words of an internal corporate memo, to "the entire fluoropolymers industry"—an industry responsible for the high-performance plastics used in kitchen appliances, computer cables, implantable medical devices, and the bearings and seals in automobiles and airplanes.

The chemical industry regulates itself. Despite the national obsessions with self-image, diet, and longevity, most Americans have no better understanding of the actual substances that most powerfully influence their biological existence than do the ciliates that crawl along the ocean floor. Each day, on average, an American man puts eighty-five man-made chemicals into his body; an American woman takes in nearly twice as many. They ingest these substances, with names resistant to memorization or even pronunciation, every time they inhale, eat, drink, bathe, or apply makeup. Many of these chemicals linger in their organs, tissues, blood. Some, like PFOA, never leave. There are more than eighty-five thousand synthetic chemicals in regular circulation. More than half a century after *Silent Spring* and forty years after Congress passed the Toxic Substances Control Act, the American government has restricted use of six of them.

What knowledge that does exist about these chemicals tends to be closely guarded, restricted to the private laboratories—at Dow Chemical, DuPont, 3M—in which they were invented. Until a 2016 amendment to TSCA, the EPA had ninety days to review the safety of each new chemical that hit the market. It did not conduct its own tests; it had to rely on independent data. Most of that data came from the firms that manufactured the chemicals. If the EPA failed to take immediate action, the chemical was deemed safe forevermore. PFOA and the tens of thousands of other chemicals introduced before the 1976 passage of TSCA were already assumed to be safe. This state of affairs could be tolerated if, as Rob Bilott had originally believed, the chemical companies were responsible stewards of the public interest. The PFOA case proved they were not. Harry Deitzler, a plaintiff's lawyer in West Virginia on Bilott's team, would say that Bilott "lifted the curtain on a whole new theater. Before that letter, corporations could rely upon the public misperception that if a chemical was dangerous, it was regulated. Rob's letter said, wait a minute—the fact that a chemical isn't regulated does not mean that it's less dangerous to public health." Bilott was showing DuPont what self-regulation really was.

Taft's clients grumbled. What side was Taft on, exactly? Bilott's colleagues asked the same question. "I'm not stupid," said Bilott, "and the people around me aren't stupid. You can't ignore the economic realities of the way the business is run." He couldn't tell to what degree his professional isolation at Taft derived from his working for the enemy or from the thousands of hours he had spent in isolation from his colleagues. But he didn't dwell on it. There were always new documents awaiting him, hundreds of thousands of documents, and never enough time.

Rob's Famous Letter led, in 2005, to DuPont's reaching a $16.5 million settlement with the EPA, which had accused the company of concealing its knowledge of PFOA's toxicity. DuPont was not required to admit liability. It was the largest civil administrative penalty the EPA had obtained in its history. The fine represented less than two percent of the profits earned by DuPont on PFOA that year.

Bilott never represented a corporate client again.

During the summer of 2001, Bilott received a phone call from a night-school teacher in Parkersburg named Joseph Kiger. The previous Halloween, Kiger had received an odd note from the local water district, folded into his monthly bill. It explained that an unregulated chemical named PFOA had been detected in the drinking water in "low concentrations." There were several other baffling lines, which Kiger underlined as he read, malleable language that appeared designed to inoculate readers against a hidden truth. It ended with the statement: "DuPont reports that it has toxicological and epidemiological data to support confidence that exposure guidelines established by DuPont are protective of human health." Tautological nonsense.

Still Kiger might have forgotten about it were it not for his wife's alarm. Darlene had already spent much of her adulthood thinking about PFOA. Her first husband, her high school sweetheart, was a chemist in DuPont's PFOA lab. They had been the model Washington Works family. "When you worked at DuPont in this town," said Darlene, "you could have everything you wanted." DuPont paid for her husband's education, secured his mortgage, and paid a generous salary. DuPont even gave him a free supply of PFOA, which Darlene used to clean the car and wash the dishes. Sometimes her husband came home with a fever and diarrhea after working in one of the PFOA storage tanks, despite wearing a protective outfit that co-workers called "the space suit," but it was an acceptable tradeoff. Getting sick from the PFOA tanks wasn't uncommon; workers called it "Teflon flu."

In 1976, after Darlene gave birth to their second child, her husband surprised her by returning home one evening in a gray DuPont uniform. He explained that he was no longer allowed to bring his work clothes home, because DuPont had found out that PFOA was causing health problems for women and birth defects in children. Darlene would remember this six years later when, at thirty-six, she had to have an emergency hysterectomy and again, eight years later, when she had a second surgery. After the letter from the water district arrived, Darlene thought of her ex-husband's clothing, her hysterectomy. What did DuPont have to do with their water?

Joe called the West Virginia Department of Natural Resources ("They treated me like I had the plague"), the Parkersburg office of the state's Department of Environmental Protection ("'nothing to worry about'"), the water division ("I got shut down"), the local health department ("just plain rude"), and even DuPont ("I was fed the biggest line of bullshit anybody could have been fed"), before a scientist in the regional EPA office finally took his call. Kiger read aloud the letter from the water district.

"Good God," said the scientist. "What the hell is that stuff doing in your water?" He sent Kiger information about the Tennant lawsuit. On the court papers Kiger kept seeing the same name: Robert Bilott, of Taft Stettinius & Hollister, in Cincinnati.

The only remedy appropriate to the scale of the crime was a lawsuit on behalf of everyone whose water was tainted by PFOA. In all ways but one, Bilott himself was the ideal candidate to lead such a suit. He understood PFOA's history as well as anyone at DuPont and understood the regulatory law as well as anyone at the EPA. But a class-action suit would be an existential threat against the industry that supported Taft's business. It would establish a precedent for suing corporations over unregulated substances. It would be taken as a declaration of all-out war. Bernard Reilly made this point explicit to Terp in another bullying phone call. Did Taft want to risk everything for the personal crusade of one stubborn partner? Terp didn't bend.

Bilott had planned to file the lawsuit on behalf of the water districts closest to Washington Works. But tests revealed that six districts and dozens of private wells were tainted with levels of PFOA as much as seven times higher than DuPont's own internal safety threshold. Seventy thousand people were drinking poisoned water. Some had been doing so for decades.

This wasn't enough, however, for a court of law. PFOA appeared on no federal or state list of contaminants. How could Bilott claim that thousands of people had been poisoned if PFOA, legally speaking, was no more harmful than water itself?

The best metric Bilott had to judge a safe exposure level was DuPont's own internal standard, one part per billion. When DuPont learned that Bilott was preparing a new lawsuit, however, it announced that it would reevaluate that figure. As in the Tennant case, DuPont formed a team composed of its own scientists and scientists from the West Virginia Department of Environmental Protection. It announced a new threshold: one hundred and fifty parts per billion.

Bilott found the figure "mind-blowing." But the state endorsed the new standard. Within two years, three of the DuPont lawyers who established the new limit were hired by the West Virginia Department of Environmental Protection. One was placed in charge of the entire agency. Bilott was shocked. His West Virginian colleagues were not. Bilott needed a new legal strategy.

For once West Virginia's legal eccentricities would come

to his defense. A year earlier, it had become one of the first states to recognize what is called, in tort law, a medical-monitoring claim. Under this claim, a plaintiff needs only to prove exposure to a *potential* toxin. If the plaintiff wins, the defendant is required to pay for regular medical testing. Should the plaintiff one day become ill, he or she can sue, ret-roactively, for damages. Though four of the six water districts contaminated with PFOA lay across the river in Ohio, Bilott filed in West Virginia.

The EPA, meanwhile, had opened its own investigation into PFOA. In 2002, the agency released its initial findings: PFOA might pose human health risks not only to those drinking tainted water but also to the general public—anyone, for instance, who had cooked with a Teflon pan. The EPA was particularly alarmed to learn that PFOA had been detected in American blood banks—something 3M and DuPont had known as early as 1976. PFOA contami-nation was no longer a regional crisis. It was a slow-motion national catastrophe. By 2003 the average concentration of PFOA in the blood of an adult American was four to five parts per billion.

In June 2004, when the EPA sued DuPont, it cited Du-Pont's failure to report its knowledge that PFOA had been detected in the public water supply. Three months later, Du-Pont settled the class action suit with Bilott. It agreed to install filtration devices in the affected water districts and pay a cash award of seventy million dollars. It would fund an epidemiological study to determine whether there was a "probable link"—a term that evaded any declaration of causation—between PFOA and any disease. If links existed,

DuPont would pay doctors to monitor the affected group in perpetuity.

A reasonable expectation, at this point, was that the lawyers would move on. "In any other class action you've ever read about," said Deitzler, "you get your ten bucks in the mail, the lawyers get paid, and the lawsuit goes away. That's what we were supposed to do." Taft had recouped its losses, and more: Bilott's team of lawyers earned $21.7 million in fees. His colleagues, for the first time, began to look at him with respect. Bilott had every reason to walk away.

He didn't.

There was, in Bilott's phrase, "a gap in the data." DuPont's internal health studies, damning as they were, were limited to employees. Those studies also had a habit of being terminated, or "put on hold," once incriminating results began to come in. DuPont could argue—it had already argued—that even if PFOA caused health problems, it was only because the test subjects were factory workers exposed to exponentially higher levels of the chemical than residents who drank tainted water. Without more data, the epidemiological panel had no illness to study. The gap in the data allowed DuPont to claim that it had done nothing wrong.

Bilott represented seventy thousand people who had been drinking PFOA-laced water for decades. What if the settlement money could be used to test them? Bilott pushed to make each class member's settlement fee contingent on a full medical examination. Within months, nearly seventy thousand West Virginians traded their blood for a four-hundred-dollar check.

The team of epidemiologists was flooded with medical data and there was nothing DuPont could do to stop it. According to the settlement, DuPont had to fund the research without limitation. The scientists had hit the epidemiological jackpot. They designed twelve studies, including one that, using sophisticated environmental modeling technology, determined exactly how much PFOA each individual class member had ingested.

With such superfluities of data and money—the studies ultimately cost DuPont thirty-three million dollars—it was assured that whatever conclusions the panel returned would be convincing. But if nothing incriminating was found, the settlement barred Bilott's clients from bringing personal-injury lawsuits. Because of the quantity of data, the panel took longer than anticipated to perform its analysis. Two years passed without any findings. Bilott waited. A third year passed. Then a fourth, a fifth, a sixth. Still the panel was quiet. Bilott waited.

It was not a peaceful wait. The settlement fees had granted a reprieve but as the years passed without resolution, Bilott found himself in an awkward position. His reputation had been ruined—had become toxic—within the clannish community of environmental defense lawyers. DuPont's lead attorney in the class-action suit refused to negotiate with him; others refused to look him in the eye or address him in court. It became obvious in Taft's balance sheets, sent monthly to the desks of Taft's partners, that no matter how successful the DuPont case ended up, it would never come close to replacing the business Bilott had lost. He soon found himself in a familiar position: hemorrhaging Taft's money, with no guarantee of reimbursement. Ultimately he would move out of Taft's headquarters to a tiny satellite office across the river, in Kentucky.

Yet he'd never been busier. The publicity generated by Teflon headlines had aroused tremendous scientific interest in synthetic perfluorinated compounds. New studies were published weekly, if not daily. Taft continued to pay consultants to interpret the findings and relay them to the court-appointed epidemiologists. Bilott counseled class members in West Virginia and Ohio and traveled frequently to Washington to attend meetings at the EPA. Meanwhile his class members kept getting cancer and dying.

Clients called, irate and desperate. When would they get relief? Among the callers was Jim Tennant. Wilbur, who for years had suffered terribly from respiratory problems—swollen throat, burning eyes, foggy vision—had died of a heart attack. Two years later, Wilbur's wife, Sandra, died of cancer. Bilott was tormented by the thought that they had died without seeing DuPont held responsible for what it had done.

Bilott remained doggedly unexpressive but the strain began to manifest itself in ways that were obvious to his wife, if not to himself. Sarah tried to tell herself that it would be over soon but she knew that cases of this magnitude went on forever. Their sons' childhoods would not go on forever. The PFOA case had become Bilott's entire world, a world far from their home, from Cincinnati—a world that seemed far from Earth. "He's a person who doesn't lose his temper," Sarah would say. "He doesn't have a ton of other outlets. I jog three miles. I scream. I was thinking you can't keep all the stress inside." His sons thought that their father's job was to sue DuPont. Later, when they thought of it, they would ask their mother, "Is Dad still working on that case?" They called him the Lorax.

One Sunday morning in the spring of 2010, while his wife and sons were at church, Bilott was taking a shower when he became dizzy. His eyes blurred. He dried off, dressed, and carried his socks to the kitchen. He could not put on his socks. His leg refused to lift. Sarah and their sons returned to find him sitting blankly at the kitchen table. His hand shook.

"My arm feels funny," he told them. Soon after he couldn't speak. He began to lose consciousness.

At the hospital it was determined that he was not experiencing a heart attack or a stroke. The Bilotts' relief gave way to anxiety as doctors—first in Cincinnati, then at the Mayo Clinic—failed to make a diagnosis. The attacks recurred periodically, bringing foggy vision, slurred speech, and paralysis on one side of his body. They struck suddenly, without warning, and their effects lasted days. The doctors hit upon medication that helped but the episodes continued; occasionally in a meeting his right leg would jerk uncontrollably or he would become incoherent. Doctors asked whether he was under heightened pressure at work. "Nothing different than normal," said Bilott. "Nothing it hadn't been for years."

The doctors never figured it out. The only way they could reach a diagnosis was to cut his head open, like one of Wilbur Tennant's cows.

In December 2011—after seven years—the epidemiologists began to release their findings. They had detected a "probable link" between PFOA and kidney cancer, testicular cancer, thyroid disease, high cholesterol, preeclampsia, and ulcerative colitis. Clients called to thank Bilott for finally slaying Goliath.

"There was," said Bilott, "relief."

More than thirty-five hundred of Bilott's clients filed personal-injury lawsuits against DuPont. These included Sandra Follett who, though her bowel problems kept her under a self-imposed house arrest, was accused of malingering by family members who worked for DuPont; Mike Smalley, whose wife refused to drink anything but water and died of kidney cancer ("I absolutely feel DuPont is responsible for the death of my wife"); and Carla Bartlett, a kidney cancer survivor who in 2015 was the first member of the class to stand trial. DuPont had nominated Bartlett to serve as one of the first cases, most likely because she had a relatively minor claim. The concentration of PFOA in her blood was low, she had been treated successfully and at minimal personal cost, she had additional risk factors, and the cancer had not returned after twenty years. DuPont's lawyers expected the result of the Bartlett trial would undermine the thousands of cases to come. She was awarded $1.6 million. In the summer of 2016, a second plaintiff, a college professor who had lost a testicle to cancer, was awarded $5.6 million; a third, a truck driver with four children who had also survived testicular cancer, received $12.5 million. In the middle of the next trial, in February 2017, DuPont folded. It settled the entire class of lawsuits for $670.7 million.

DuPont ceased production and use of PFOA in 2013. The five other companies in the world that used PFOA ended production soon thereafter. DuPont, during its merger with Dow Chemical, severed its chemical business, spinning it off into a new corporation with the Huxleyan name of Chemours (a portmanteau of "chemistry" and "Nemours," of E. I. du Pont de Nemours). The new company replaced PFOA with similar fluorine-based compounds designed to biodegrade more quickly. Like PFOA, these new substances have not come under federal regulation. When asked about their safety, Dan Turner, DuPont's head of global media relations, said, "Extensive data has been developed, demonstrating that these alternatives are much more rapidly eliminated from the body than PFOA." More rapidly, that is, than the decades it takes the human body to flush out PFOA.

In 2016, the EPA issued a "health advisory" for PFOA and a related compound, perfluorooctane sulfonate, or PFOS. The agency did not bother to notify Bilott directly; he found out through a Google alert. The EPA warned against long-term exposure to drinking water with concentrations of PFOA or PFOS higher than 0.07 parts per billion, a guideline that Bilott considered far too lenient. The Environmental Working Group had proposed a limit of 0.001 parts per billion, though Bilott believed that anything above zero was dangerous.

An EPA health advisory is not a regulation; it is nonbinding and nonenforceable. It does, however, require a public water system to notify its customers and explain the dangers it poses. The shame of the admission is usually sufficient to force health officials to see the light. Within hours of the announcement, West Virginia's Bureau for Public Health ordered a "do not drink" advisory for the water in three communities: Parkersburg, the neighboring town of Vienna, and Martinsburg, four hours east. The state's National Guard sent convoys of tankers containing drinking water. *The Parkersburg News and Sentinel*, which for a decade had largely managed to avoid covering the class-action lawsuit facing the city's main employer—a suit in which many of its readers were plaintiffs—published an editorial titled "Don't Panic," accusing the EPA of changing its mind about PFOA "overnight." The editors noted that water in the area had tested at these levels for "decades," as if this were cause for reassurance.

In 2019, the EPA announced an "Action Plan" to address PFOA and PFOS—a plan that lacked deadlines, specific targets, and any promises of accountability. It failed to address any of the thousands of PFOA knockoffs that now suffuse the marketplace.

Actual regulations or bans, should they one day be announced, might be a source of comfort to future generations. But if you are reading this during the first quarter of the twenty-first century, you already have PFOA in your blood. It is in your parents' blood, your children's blood, your lover's blood. How did it get there? Through the air, through your diet, through your use of nonstick cookware, through your umbilical cord. Where scientists have tested for the presence of PFOA, they have found it. PFOA is in the blood or vital organs of Atlantic salmon, swordfish, striped mullet, gray seals, common cormorants, Alaskan polar bears, brown pelicans, sea turtles, bald eagles, California sea lions, and Laysan albatrosses on Sand Island, a wildlife refuge on Midway Atoll, in the middle of the North Pacific Ocean, about halfway between North America and Asia.

"We see a situation," said Joe Kiger, "that has gone from Washington Works, to statewide, to the United States, and now it's everywhere. It's global. We've taken the cap off something here. But it's not just DuPont. Good God. We have no idea what we're taking."

Bilott didn't regret fighting DuPont for twenty years, nor letting PFOA become his life's work. But he was still angry. "The thought that DuPont could get away with this for this long," he said, his tone landing halfway between wonder and rage, "that they could keep making a profit off it, then get the agreement of the governmental agencies to slowly phase it out, only to replace it with an alternative with unknown human effects—we told the agencies about this in 2001, and they've essentially done nothing. That's years of this stuff continuing to be in the drinking water all over the country. And in the meantime, DuPont fights everyone who has been injured by it."

They fought, because they were bound to. Their allegiance was not to public health, or better living through chemistry, but to shareholders. The manufacturers' strategy belonged to the *dojo kun* of "Too Big to Fail." If everybody was a carrier of PFOA, only those who suffered from blast-zone levels of exposure could be considered poisoned. Americans, on average, carried in their blood concentrations of PFOA of about two parts per billion, and three times that amount of PFOS. No concentration is safe, but two parts per billion has been normalized. Two parts per billion is our biological inheritance.

This realization led Bilott to propose a new case to his partners at Taft. It would be another class action, only this time the class would have three hundred and twenty-eight million members. He argued that the manufacturers of PFOA and its thousands of chemical cousins had been treating American citizens like the cows at Dry Run, stowing toxic effluent in their blood for decades without their permission or awareness. Bilott filed his lawsuit in Columbus, Ohio, on October 4, 2018. He expected the fight would last years, perhaps the rest of his life. If you are a U.S. citizen alive in the twenty-first century—or any other biological organism hosting a payload of synthetic poison—Rob Bilott represents you.

2

THE WASTING

Allison Gong was a marine biologist so she understood perfectly well that a sea star has no blood, brain, or central nervous system. Still she couldn't help thinking of the stars in her lab as pets. "Because of my weird personality," she said, "I form an emotional attachment, even though obviously they can't reciprocate."

The attachment deepened during her first twenty years at the Long Marine Laboratory at the University of California, Santa Cruz, where she exhibited the stars to undergraduates in her marine biology classes. (One of her first lessons: "Starfish" is a misnomer, because stars are not fish; phylogenetically speaking, the animals are asteroids.) Gong had fifteen stars in her care: eight bat stars, five ochres, a leather, and a rainbow. Each morning she greeted her menagerie with a cheerful "Hey, guys!" She checked to make sure her wards were behaving; if stars were climbing out of their tank, she'd prod them back into the water, with a gentle reprimand: "Guys! You know you need to get back in there." She recorded the temperature of the water, which was piped in from the shallows of Terrace Point, the reef on which Long Marine rested; from the laboratory's windows it was common to see cresting dolphins, back-paddling sea lions, and breaching humpback whales. For breakfast, Gong fed the stars frozen squid or lake smelt that she diced into small, digestible morsels. None of the stars, which typically live about thirty-five years in the wild and can survive more than three times as long in captivity, had ever died, at least not of natural causes. Years earlier Gong had accidentally dropped a tank on a star, crushing it. She still felt bad about it.

Gong was unprepared, then, for the discovery she made during Labor Day weekend in 2013. No sooner had she greeted her charges ("Hey, guys!") than she realized that somebody had died. The bat stars, aggressive scavengers, had glommed together in a single ball—an ominous sign. Gong peeled them off, one by one, until she found what they had been consuming: the corpse of an ochre sea star, their roommate of five years.

Two days later she noticed that some of the other stars looked poorly. Their arms were twisted around their stomachs, as if hugging themselves. Healthy stars, especially ochres, have a rough texture and a muscular consistency. These looked like deflating party balloons. It soon got to the point where she was terrified to open the door. The next day an ashen lab assistant reported that one of the stars had lost an arm. When Gong returned the day after that, the water table looked "like an asteroid battlefield." The stars were squishy and pockmarked with pullulating white lesions. Sometimes their guts spilled out of the lesions. Detached arms crawled, disembodied, around the tank.

It is not uncommon for sea stars to shed their arms in times of stress. When a curious child lifts a star out of a tide pool by one of its arms, the star may jettison the limb to escape and regenerate it later. But Gong knew that this was different. Her stars weren't shedding their arms. They were tearing them off. They were tearing them off the way a man pinned under a fallen boulder, lacking access to a sharp tool, might: by using one arm to wrench the other out of its socket. They were ripping themselves to pieces.

It seemed at first that the sickness affected only the ochre stars. But soon the rainbow star began showing symptoms. Gong arrived one morning to find it wrenching off one of its arms. She left the lab to feed other animals and by the time she returned, forty minutes later, it had ripped off two more. The leather star and the last of the ochres liquefied a few days later. The bat stars did not, however, appear to be affected—at least not negatively. For them, the mass death of their table-mates was a bonanza. They gorged on the corpses.

Gong called it "the stuff of nightmares." She had never seen anything like it. "The sheer horror and the magnitude of the self-mutilation—ripping their own arms off, the disembodied parts crawling around . . . I'd seen animals die, but it's just a one-off. Something dies, and you get on with your life. But there was no getting on."

She inquired next door at UCSC's aquarium, which also drew its water from Terrace Point. The aquarists had noticed mysterious signs of disease in their own collection, which included a pair of sunflower stars, one of the largest star species in the world. A sunflower can have as many as twenty-four slender limbs, each extending the length of a meter; it crawls in the quick, lurid manner of Thing Addams. Before long the sunflowers were losing their arms too. They were so large that when they lost their arms, they looked like farm animals mid-butchery. The aquarists hastily removed them from their tanks so that the visiting children wouldn't scream.

In his closet-sized office several yards away, Pete Raimondi had begun to suspect that whatever was killing the sea stars was not confined to Terrace Point.

Raimondi, of UCSC's Ecology and Evolutionary Biology Department, had recently found himself undergoing an unexpected career shift. He had been thrust into the role of asteroid detective. As a marine biologist who divided his time between analyzing data and conducting research trips along the Pacific coast, Raimondi was not entirely ill-suited to the part. There was a private-eye quality to his round, inquiring face, active eyes, and impatient manner. He tended to wear sandals and cargo shorts in the lab but with a fedora and distressed suit he could pass for Jake Gittes. Like many in his field, Raimondi had found that his research had shifted from data analysis to something closer to criminal investigation, as the ecosystems he studied had come to resemble murder scenes.

Raimondi knew more than anybody alive about the condition of sea stars along the Pacific coast. For a decade he had served as the principal investigator at MARINe, the Multi-Agency Rocky Intertidal Network, a name clumsy enough to disguise one of the most ambitious efforts to catalog the vanishing world. Every year, MARINe sent researchers to two hundred sites along the Pacific coast between Graves Harbor, on the Alaska Panhandle, and Punta Abreojos, in Baja, Mexico. They took censuses and recorded detailed observations about more than a thousand species, including at least fifteen sea stars. The online database was open to the public. It was designed to establish a baseline of environmental conditions so that when something unusual was detected, it could be quantified—to determine whether an event was insignificant, alarming, or apocalyptic. MARINe was the first such comprehensive monitoring system in the United States; there was only one other comparable system in the world, tracking the Great Barrier Reef. Scientists did not understand the distribution of marine life with any great accuracy. The oceans remained a chaos. It was understood that human activity was altering their composition in dramatic ways, but the ramifications of these changes were poorly measured. This was troubling because if the composition of the oceans changed too drastically, sea stars would be the least of our concerns. To understand the future health of human civilization, you need only to consult the health of the oceans, on which the health of our species depends.

In the spring of 2013, Raimondi began receiving reports of high levels of sea star wasting syndrome. "Wasting" is a generic term that describes various forms of physical deterioration, which in the case of sea stars includes splotchiness, sores, deflation, and the jettisoning of limbs. Any number of insults, both environmental and pathogenic, can lead to wasting. It's the asteroidal equivalent of coming down with a severe case of the flu, afflicting about one percent of stars at any given time. But when a high percentage of animals succumb, it means something is wrong. It is the difference between a stubborn cough and an epidemic.

Raimondi quickly concluded that it was at least an epidemic, if not something much worse. In March, a marine-water-quality specialist at the University of Washington reported signs of wasting in every sunflower star found on the coast of Vashon Island. In April, a technician at Oregon State noticed wasting symptoms in ochre stars at the Carl G. Washburne Memorial State Park. In June, researchers observed wasting stars at Sokol Point on Washington's Olympic Peninsula. In August, Raimondi himself encountered wasting ochres on a research trip in the Gulf of Alaska to Kayak Island, sixty miles from the nearest town. Then the reports got weirder.

During the fall, sightings increased in number and virulence. The geographic range of the events was startling. Mottled stars died at the Anchorage Museum and Pacific blood stars died at Point Loma in San Diego. The Seattle Aquarium's veterinarian, panicked at the sight of the sick stars, had them quarantined and doused with antibiotics; when that failed, she euthanized every star that showed signs of illness. The sea star population of Terrace Point, in Long Marine's backyard, crashed. Divers observed wasting stars in sub-tidal reefs and crabbers found wasting stars in pots pulled up from depths of three hundred feet. Nobody knew what to call it. Was it a die-off? A plague? A population crash? An extinction event? Scientists began to call it "the Wasting."

Gong had never witnessed anything like it but Raimondi had. In 1982, while a graduate student at UC Santa Barbara,

Raimondi observed firsthand the effects of the strongest El Niño episode of the twentieth century. Temperatures in the Pacific Ocean spiked by as much as ten degrees Fahrenheit. Sea stars, among other affected marine creatures, died in shocking quantities from wasting syndrome. This occurred again after the 1997–98 El Niño, when some asteroid populations collapsed by more than half. In these incidents warm water seemed to be the common variable, particularly since localized wasting events had often occurred in Southern California during warmer-than-average years. Temperature rise was believed to have contributed to other marine die-offs: the sudden collapse of the Long Island Sound lobster fishery in 1999; the mass bleaching of Caribbean coral reefs in 2010; the death of thousands of pelicans on the beaches of northern Peru in 2012.

Yet warmer water didn't appear to be responsible for the Wasting. Although the oceans were warming globally, the water along the Pacific Coast had been relatively cool for the dozen years before the Wasting, and the first deaths were observed in the colder waters of the Pacific Northwest. Raimondi had not heard of a wasting event in Alaska before. What surprised him most was the speed of death: the siege was sudden, total, and seemed not to discriminate by species. Never had Raimondi seen disembodied arms walking around. Or sunflowers "exploding." Nor had he seen a ghost star. Wasting tends to be a gradual process, as the star deteriorates over the course of days or weeks. But the Wasting attacked with such ferocity that some of the stars rotted instantly, without being able to take another step. Their soft tissue dissolved and decomposed, devoured by fuzzy white bacteria that resembled mold on a loaf of bread, while the star's hard white spicules—their spines, which are composed of calcium carbonate—remained. This left behind a ghostly impression of the star; it was, quite literally, a chalk outline.

Raimondi described the ghost stars as "creepy," a term biologists tended to avoid. The Wasting had that effect. It made scientists, who tend to choose their words with severe caution, speak like teenagers. In conversations about the Wasting they kept using words like "horror," "shock," and "nightmare."

Biologists investigate a mortality event the way that the CDC tracks a viral outbreak or an FBI special agent tracks a highway killer. It's not enough to identify the victims. They need to determine the sequence in which they died. They have to trace the violence back to its source. No matter what angle he took, however, Raimondi couldn't detect a pattern. In the intertidal region alone, the mortality rate averaged about seventy-five percent. But the stars died at different rates. Some became ghost stars within hours, some took a week to succumb, and still others managed, miraculously, to recover. If the epidemic was influenced by warm water, why had it gotten worse during the winter? If it was caused by pollution, why was it occurring everywhere? If it was caused by a pathogen, why didn't it appear to spread outward from some point of origination, instead of leapfrogging about the coast? In the middle of some of the most devastated regions, they found pockets of healthy stars. In unaffected areas, they found pockets of wasting. There was no sense to it. Raimondi even began to question whether the Wasting was an example

of wasting syndrome. Perhaps it was something different altogether, something entirely unprecedented.

Cameramen from CBS, NBC, and CNN began trailing Raimondi on his research expeditions. Boats of seasick journalists appeared in the bay. The British tabloids ran articles
with headlines like MYSTERY AS MILLIONS OF STARFISH WASH
UP DEAD ON WEST COAST and MYSTERIOUS PLAGUE CAUSES
STARFISH TO RIP OFF THEIR ARMS—AND SCIENTISTS DON'T
KNOW WHY. The press attention had its benefits. Alarmed citizens began to mobilize, independently and then in factions,
to save the stars. Logging in to an interactive Sea Star Wasting
map that Raimondi created, they swarmed the Pacific coast,
including sites that researchers never visited, tallying their
observations. As the data accumulated—deliquescing stars
were detected as far away as the northern Atlantic coast—
Raimondi's Wasting map rapidly populated with markers of
death. The Wasting had become, by Raimondi's estimation,
"the best-reported marine disease ever." But still no coherent
pattern emerged.

Amateurs wrote him with their theories. Many blamed
global warming and its attendant marine catastrophe, acidification, which occurs as oceans absorb higher levels of carbon dioxide. A particularly determined, conspiratorial set
blamed the Fukushima nuclear crisis. Others blamed power
lines along the coast for bombarding reefs with electromag-
netic radiation. One man claimed that the plague had been
caused by Christmas. He proposed that fir trees, farmed

in Alaska, carried with them a bacteria lethal to sea stars, which they shed into the water from tanker ships in transit to Southern California.

Raimondi's citizen researchers were motivated less by the spirit of scientific inquiry than sheer panic. Donna Pomeroy, a retired wildlife biologist who for two decades had lived across the street from Pillar Point reef in San Mateo County, was sickened to see the stars, which normally clung to rock ledges, peel off and collapse in the sand. "They looked as if they were made of wax and had been left too close to a heat lamp," she said. "The arms were literally dripping off." She also began to notice, in startling abundance, a Pepto-Bismol-colored mollusk called the Hopkins' rose nudibranch. Though it was common to go years without spotting one, hundreds of them now took over the reef. The writer Mary Ellen Hannibal, who participated in the star surveys, compared it to walking into a redwood forest and finding candy canes growing out of the branches. "The nudibranchs are gorgeous," said Pomeroy, "but it's scary to see these changes happen so fast. There's a bigger picture, and we don't know what it is."

Local science teachers recruited students to catalog the destruction. Catherine Lyche, a junior at Monterey's Santa Catalina School, had an emotional attachment to sea stars; she would "scream with joy" when she found them on school tide-pooling trips. So she was disturbed to find stars that were wrinkled, armless, and decomposing. "Even my teacher didn't know what was causing it," said Lyche. "That was troubling." Her classmate Katie Ridgway was startled when she couldn't find any stars on a trip to the local Asilomar reef. A year earlier the stars had been everywhere. "It was like, wow," she said. "Did *I* do something to cause this?" On a school break she returned to Seattle, where she had grown up, and found that the Puget Sound reef she used to explore as a child was also starless. "It made me wonder, if this continues, and water rises, and another virus affects another organism, what's going to happen by the time I have kids?"

"If we're not able to predict something as significant as this," said Lyche, "what else don't we see coming?"

In the following years Raimondi and his colleagues began to conclude that no single factor was responsible for the Wasting. It might have been facilitated by some obscure permutation of pathogens and variations in environmental conditions, including storms, ocean currents, salinity, industrial pollution, and rising temperatures—or none of these. It could not even be classified as a disease, because there was no consistent set of symptoms. If anything it was a "syndrome": a collection of disorders that happen to occur at the same time. Given the quick onset of attack and the scale of its destruction, Raimondi still believed there was some unifying cause. But he could not determine what it was.

He found himself in the position of an investigator who has intimate knowledge of his suspect—the killer's tendencies, eccentricities, and modus operandi—who knows everything about the suspect, that is, except his identity. The uncertainty forced Raimondi to take a more philosophical attitude; he began to sound less like a private eye and more like the Wasco prophet who, after seeing visions of the future on the eve of the colonizers' arrival, told his tribe, "Soon all sorts of strange things will come. No longer will things be as before. You people must be careful."

On a winter afternoon a member of Raimondi's team named Melissa Redfield led a group of amateur tidepoolers ten minutes south to Natural Bridges State Beach. Below low vertical cliffs, a margin of flat sand yielded to long platforms of Santa Cruz mudstone. They were slippery with algae and soft enough that sea urchins could burrow cubbies into them. Two children, trailed by their mother, examined the reef's shallow wells for life. They yelped every time they spotted a hermit crab, a purple sea urchin, a sea anemone with neon-green tentacles. A family of Japanese tourists did the same. A solitary woman kneeled facing the ocean and played a dirge on a recorder.

The citizen scientists, warming to their despair, complained that they could not find any sea stars. But Redfield began to spot them almost immediately. Near the edge of the tide she abruptly belly flopped on the mudstone, craning her neck to peer into a jagged fracture. Though she shined a flashlight into the baseball-sized cavity where an ochre star was hiding, it took a full minute for her guests to see it, so expertly had it camouflaged. She soon found another star, then many more. Most were violet or purplish pink, hiding in holes and pockmarks and, in one case, beneath a sea urchin. After half an hour, she had found about a dozen ochre stars. Most were the size of a quarter or smaller; the largest was the size of an adult hand. All appeared healthy except for one of the larger stars. It had lost an arm, and at the base of another a white lesion was forming.

This was not atypical but really there was no typical. The

recovery in the years following the Wasting was as baffling as the event itself. At many locations, stars had disappeared permanently. At others, amputee stars began to regrow their arms, and populations seemed to be recovering, only to crash belatedly, months later. For a while it seemed that where star populations had not entirely crashed, smaller individuals dominated; Rich Mooi, the curator of invertebrate zoology and geology at the California Academy of Sciences, compared it to a forest fire: "The forest burns down, and then the seedlings come." But larger individuals also began to reappear. The forest fire analogy was imperfect anyway, because many of the smaller stars detected were not newborns. Sea stars grow slowly; by the time they grow large enough to be observed, they are likely several years old. The stars at Natural Bridges were not babies but survivors.

"It's hard for me to think of the big picture," said Redfield, who was happy to see any stars at all. "I don't want to think about the big picture."

The big picture was that the cascading consequences of the Wasting were as various and inextricable as its causes. Sea stars eat sea urchins; unchecked by its main predator, sea urchin populations exploded to one hundred times pre-Wasting levels. Sea urchins devour kelp. After the Wasting, urchins reduced the kelp canopy along the Pacific coast by more than ninety percent, turning kelp forests, which provide nutrients and shelter to a diverse pageant of sea life, into kelp deserts. This led to the phenomenon known as urchin barrens: marine wastelands devoid of life apart from a carpet of purple-spiked urchins. Having devoured the kelp, the urchins began to starve, leading to another mass die-off, and the boom-bust cycle charged ahead toward some unrecognizable, if no longer unimaginable, future.

Raimondi speculated that the high concentration of stars along the coast that tidepoolers took for granted might not be a natural state of affairs but an aberration, caused by vast changes to the intertidal ecosystem during the previous century. If the population of stars had wildly increased from historic levels, it could have become more susceptible to collapse. Seen from this vantage, the Wasting might be a natural corrective to human-aided overpopulation.

Yet as the years passed, there were indications that the stars were correcting back. Ochre stars north of San Francisco Bay were found to have undergone rapid genetic changes during the Wasting; other species experienced similar shifts. Under enormous stress, the stars had evolved before researchers' eyes. It seemed possible that the sudden evolutionary leaps would be permanent.

Raimondi viewed such developments with cool professional distance. The younger scientists and citizen volunteers, however, took it personally. They had been traumatized by observing, in their own lifetimes, countless extinction events and environmental calamities unprecedented in the history of civilization. They understood that most of the dying species went extinct without notice, far from civilized society, their extinctions noted only belatedly, including many that disappeared before scientists could identify them in the first place. The notion that the Wasting might presage some decisive, profound transformation of the oceans did not seem to them far-fetched.

"Raimondi sees it as a big experiment," said Jan Freiwald, a marine ecologist who led an organization that trained amateur divers to conduct biological surveys. "He takes himself out of it. But we just don't know how big the effects might be. The worst thing is when you see the other sea stars eating one that is wasting. You think, *No, don't do it!*"

"It makes you feel sad," said David Horwich, a volunteer diver who was one of the first to detect the Wasting. "Is it a onetime event or a harbinger of worse to come?"

"It feels apocalyptic," said Mary Ellen Hannibal. "Whatever is going on with the sea stars has the sense of an immersive event that's not visible to the eye, that's pulling species out from underneath."

All that Raimondi could do was continue to monitor the stars. He watched to see whether their ecosystems recovered and he looked for genetic indicators that might predict what would happen in future wasting events. His work relied on the vast network of citizen volunteers driven by the crisis to action. They generated an enormous amount of data. The only shortcoming of this approach was that amateur scientists struggled to detect the juvenile stars, which can be smaller than a pinkie fingernail.

For this reason some of the most valuable sightings had come from children as young as three years old. Preschoolers made shrewd detectives. Unlike their parents, they loved getting dirty and crawling along the reef, peeking below ledges and reaching into dark recesses. They had excellent eyesight, boundless curiosity, inexhaustible energy. They were deeply persistent. They enjoyed solving problems. They liked to feel that they held the fate of the world in their hands.

3

HERE COME THE WARM JETS

"It just seems like a beautiful day in Southern California."

It was late January in Porter Ranch, a hillside neighborhood of gouty model homes straddling the Ronald Reagan Freeway on the northern fringe of Los Angeles. Bryan Caforio, a young trial lawyer campaigning for Congress, gestured at the pink and orange striations of sky above Aliso Canyon. "It seems like a beautiful sunset in a wonderful community."

It did seem so. Caforio sat at an outdoor table at the Starbucks in Porter Ranch Town Center, overlooking an oceanic parking lot vibrating with shoppers. The air was dry, still, 70.3 degrees. The canyon's foothills bronzed in the falling daylight. But there were clues, hidden in plain sight, that suggested the presence of an alien life-form.

Parked in front of the Starbucks was a trio of news vans, antenna masts projecting from their roofs. A cameraman emerged, stared quizzically up at the canyon, and reentered the van. Next to the Supercuts, security guards stood before a nondescript storefront; stenciled on the window were the

words "Community Resource Center" and, in smaller letters, "SoCalGas." The guards demanded identification and
dismissed anyone who tried to take a photograph. At the
entrance to Bath & Body Works, a device that resembled an
electronic parking meter balanced on a tripod; the digital
display, visible to anyone who looked at it (though only one
person bothered to look at it), read BENZENE, followed by a series of indecipherable ideograms. Among the rows of parked
cars was a preponderance of silver Honda Civics bearing the
decal of the South Coast Air Quality Management District.
Inside the cars, men sat in silence, waiting.

Beyond the Walmart and the Ralphs rose a tiered slope
of beige houses with orange clay-tiled roofs and three-car
garages. The lawns were tidily landscaped with lavender,
succulents, hedges of rosemary, cactuses, and kumquat
trees. In the backyards: swimming pools, hot tubs, grills,
fire pits, fishponds. The meandering streets were desolate,
apart from the occasional unmarked white van. As you ascended the canyon, reaching gated "villages" with names
like Renaissance and Promenade, the police presence increased. On Sesnon Boulevard, the neighborhood's northern boundary, an electric billboard propped in the middle
lane blinked: REPORT CRIME ACTIVITY; LAPD IN THE AREA;
CALL 911. Holleigh Bernson Memorial Park was empty aside
from three cop cars, patrol lights flashing. The vans, empty
winding streets, and police cruisers recalled the paranoid
final scenes of E.T., when Elliott and his friends are chased
around the suburbs before taking flight over the roofs, into
the canyon—and not just because Steven Spielberg shot
E.T. in Porter Ranch. Several blocks away, E.T. Park was also
empty.

Beyond the last homes, spaced along the canyon ridge, stood a graffiti of spindly metal structures that resembled antennas or construction scaffolding or alien glyphs. Until about a month earlier, most residents of Porter Ranch had ignored them.

"You look at the hills, you see a few towers," said Caforio. "But do you really know what they are?" He shook his head. "You try to say, 'Hey, we're having an environmental disaster right now!' But it just looks like a beautiful sunset."

Anyone who happened to glance farther up, several hundred feet above the ridge, would have noticed another unusual sight that week: a single-engine Mooney TLS airplane, which looks like something Cary Grant might have flown in *Only Angels Have Wings*, flying loops around the mountains. Its pilot, Stephen Conley, was the first person to grasp the scale of what was happening in Porter Ranch. He took his initial flight after the Southern California Gas Company filed a terse report with the California Public Utilities Commission on October 25, 2015, noting that a leak had been detected at a well in its Aliso Canyon storage facility. Under "Summary," the report read, "No ignition, no injury. No media."

The local news media did begin to take notice, however, after Porter Ranch residents complained of suffocating gas fumes. In response, SoCalGas released a statement pointing out that the well was "outdoors at an isolated area of our mountain facility over a mile away from and more than 1,200 feet higher than homes or public areas." It assured the public that the leak did not present a threat.

Timothy O'Connor, the director of the Environmental Defense Fund's California Oil and Gas Program, had read about the complaints. But he had not thought much of them until a week later when, at a climate-policy event in downtown Los Angeles, he heard that SoCalGas was flying in experts from around the country to help plug the leak. When he returned home O'Connor opened his laptop and tried to learn everything he could about Aliso Canyon. How much gas was stored inside those mountains? How pressurized was it? He was trying to figure out, in other words, the answer to one question: How much gas could leak out?

The foothills of the Santa Susana Mountains on which Porter Ranch was built, O'Connor learned, once belonged to J. Paul Getty. His Tide Water Associated Oil Company hit crude in 1938 and did not sell the land until the early 1970s, after it had extracted the last drop. The oil field was bought by Pacific Lighting, which used the vacated cavity to store a vast subterranean sea of natural gas. Formed in the Miocene epoch, the cavity began more than seven thousand feet below the surface. With a capacity of eighty-four billion standard cubic feet, it was one of the largest natural gas reservoirs in the country. The facility functioned as a kind of gas treasury. When prices were low, the company hoarded the gas; when prices were high, it released the gas into pipelines that snaked through Los Angeles, providing power to solar- and wind-energy facilities, fueling stoves, and heating homes.

The one hundred and fifteen wells in Aliso Canyon could be envisioned as long straws dipping into a vast subterranean sea of methane. The leaking well, SS-25, was cannulated by a steel tube seven inches in diameter that descended 8,749 feet from the ridge. Its superstructure, which a layperson would struggle to distinguish from an oil derrick, was plainly visible from many of the streets in Porter Ranch. When residents looked at it through a pair of binoculars, they could make out, flying from its highest girder, an American flag.

After conducting some basic calculations, O'Connor arrived at a shocking conclusion. The canyon was like an overinflated balloon: given the pressure and quantity of gas stored within, a puncture could release in a single day as much gas as nearly two thousand houses would consume in a year. As it turned out, O'Connor was mistaken—the figure ended up being much higher than that—but he included it in an urgent letter he sent the next day to the governor's senior energy adviser and members of the state's major regulatory agencies. That evening O'Connor attended a hearing at the Porter Ranch Community School. One hundred panicked residents complained that the gas fumes had caused them headaches, respiratory problems, nosebleeds, and vomiting. No state official attended. The next morning, having yet to receive a response to his letter, O'Connor realized that he didn't have to wait for the state to take action. He could call Stephen Conley.

At 10:30 a.m. on November 5, Conley took off from Lincoln Regional Airport, just north of Sacramento. An atmospheric scientist at the University of California in Davis, Conley ran a freelance air-monitoring business, hired by public agencies and climate watchdogs to loop over oil and gas fields, measuring methane concentrations with a device called a Picarro analyzer. Timothy O'Connor asked Conley to fly circles around the methane plume for fifteen minutes, and notified SoCalGas as a courtesy. O'Connor began to receive a flurry of angry text messages from the utility's executives. They all delivered the same message: the flight was "inappropriate" and unsafe.

This mystified O'Connor. There were no concerns about explosions or air visibility, so what would be unsafe about the flight? It took some time for him to realize that SoCalGas was not concerned for the safety of the *pilot*. The executives claimed to be concerned for their workers, operating cranes and drills in the effort to plug the leak: they might become "distracted" by the sight of an airplane. O'Connor found this reasoning odd, because Porter Ranch lay in the flight path of Van Nuys Airport. Nearly six hundred flights passed overhead every day. He proposed that the airplane keep one mile away from the well site. SoCalGas said even that would be unsafe. O'Connor asked whether there was a distance from the well at which the airplane could fly. The company said there was not. When Conley was less than twenty minutes from Aliso Canyon, O'Connor canceled the flight.

Two days later Conley was back in the air, this time on assignment for the California Energy Commission. Conley flew alone; though the Mooney has a second seat it was rarely occupied because the extreme turbulence of his flights usually made his passengers throw up. Conley was used to the turbulence but the sight of the Santa Susana Mountains made him sick. The southern slope of the ridge, hidden from Porter Ranch, was brown and dead, the chaparral depilated, the surface scarred by access roads, the gas and oil structures littered about the slope like crushed beer cans.

The normal concentration of methane in the atmosphere is two parts per million. Conley's Picarro registered fifty parts per million. He figured something was wrong with the instrument but a backup analyzer gave the same reading, over and over. "Oh my God," Conley said to himself. "This is real."

What is real to a climate scientist is abstract to the rest of us. The study of climate is a study of invisible gases. To translate findings to a public lacking a basic understanding of atmospheric chemistry, scientists must resort to metaphor and allegory. They must become writers, publicists, politicians. This tends not to come easily to them. The leak at Aliso Canyon, Conley discovered, was the largest methane leak in American history. But what did that mean?

You could begin by comparing emissions from the gas leak at Aliso Canyon with other pollution sites. Conley had logged about fifteen hundred hours of flight time over oil and gas fields, moonscapes like the Julesburg Basin in Colorado and the Bakken Formation in North Dakota. The highest methane-emission rate he had ever recorded was three metric tons per hour. The methane was leaking from Aliso Canyon at a rate of forty-four metric tons per hour. By Thanksgiving, it had increased to fifty-eight metric tons per hour, doubling the rate of methane emissions for the entire Los Angeles basin. This fact takes some effort to absorb. It means that the steel straw seven inches in diameter plugged into Aliso Canyon was by itself producing twice the emissions of every power plant, oil and gas facility, airport, smokestack, and tailpipe in all of Greater Los Angeles combined.

One hundred and nine thousand metric tons of methane escaped the Aliso Canyon well in total. Over a twenty-year period, methane is estimated to have a warming effect on Earth's atmosphere eighty-four times that of carbon dioxide. By that metric, the Aliso Canyon leak produced the same amount of global warming as 1,948,085 gasoline cars do in a full year. During the four months that it lasted, the leak contributed roughly the same amount of warming as the carbon emissions produced by the entire state of Connecticut. If well SS-25 were its own country, it would have contributed to global climate change at a rate exceeding that of Senegal, Uruguay, Bolivia, Laos, Latvia, Lithuania, Zimbabwe, Albania, Nicaragua, Panama, Jamaica, Georgia, Costa Rica, Honduras, Slovenia, Armenia, Iceland, Jordan, Yemen, Kenya, and Ethiopia. SS-25 would rank between Lebanon and Syria.

These facts, despite their world-historical significance, still failed to make much of an impression locally and nationally, let alone internationally. What was one more airborne toxic event at a time when the global climate was itself an airborne toxic event? As gargantuan as the Aliso Canyon emissions were, their influence on the climate would have no direct effect on Porter Ranch. Residents were as concerned about the leak's contribution to atmospheric warming in the years and centuries to come as everyone else on the planet—which was to say, not especially. The planet already bathed in leaking invisible gases that would lead to consequences too overwhelming to imagine. What difference was another Lebanon's worth of emissions?

The inhabitants of Porter Ranch were terrified, however, about what the gas might do to their brains and their lungs.

In the tense months that followed the leak, some residents found the smell of gas so overwhelming that they sealed their windows and doors and refused to go outside. Others could not smell the gas and experienced no symptoms. Sometimes those with severe symptoms and those without lived in the same household. In the absence of reliable information from SoCalGas or the state, Porter Ranchers underwent their own transformation: They became amateur chemists, epidemiologists, sociologists, political scientists, detectives. They began to develop their own hypotheses.

Charles Chow, for instance, believed that "yellow spots" were "coming out of the atmosphere." Chow was a seventy-six-year-old retiree with mirthful eyes and springy joints. Forty-seven years earlier, Charles and his wife, Liz, were among the first to buy property after the horse ranches were cleared for suburban tracts. He spent hours in his driveway each day caring for a 1992 burgundy Cadillac Brougham Elegante and a 1986 Rolls-Royce Silver Spirit. Across the street, which was called Thunderbird Avenue, he parked his 2002 black Thunderbird.

Chow had noticed the spots while installing new shocks on his Brougham. Each spot was about the size of a yellow split pea. He started seeing them everywhere: on the windshields, on the Brougham's vinyl roof, on his home's canyon-facing picture windows. He knew that whatever was causing the spots was getting worse when Chaka Khan, his Chihuahua–miniature pinscher, began suffering severe respiratory problems. Next Liz began suffering chronic headaches, eye irritation, and a sore throat. Her doctor said there was nothing he could prescribe her. The only remedy, he said, was to leave Porter Ranch. Most of their neighbors had already gone. The Chows stayed but they drove four times each month to a vacation rental on the Baja peninsula, sixty miles south of the border, "just to get out of the atmosphere." In the Mexican air, Liz's symptoms vanished.

While Chow hunted yellow spots, he was joined in his driveway by Rick Goode, a neighbor of twenty-five years with a wiry build and birdlike gait. Goode wanted Chow's advice about legal representation. Two dozen plaintiff's firms had descended on Porter Ranch, competing to sign as many clients as possible. What did Chow think of Robert Kennedy? Or Weitz & Luxenberg, which had sent Erin Brockovich to solicit clients? The previous week, Brockovich told reporters that she "started feeling kind of dizzy" within ten minutes of arriving in Porter Ranch. Chow ruled her out.

"You don't get sick that fast," he said.

"I've been having terrible headaches," said Goode. "Have you?"

"My wife has headaches every day, sore throats," said Chow. "I don't. Everybody is different."

Liz returned from a doctor's appointment. She removed her sunglasses to reveal a new cyst on her eyelid. She searched for a word to describe her general condition since October. "A malaise," she said finally.

"You felt like you were lazy," said Barbara Weiler, a woman in her sixties who was walking her dog very slowly several blocks away. She first experienced the malaise in gym class. "It was obvious when we were using the resistance bands. We felt like we didn't want to work as much as we normally would."

Vicki, an apparel designer who refused to give her last name out of fear that something she said could be used against her in a future lawsuit, stood paces from her front step in large Prada sunglasses and a pinkish turtleneck sweater beside Joey, her nine-year-old Australian shepherd–papillon rescue dog. Years earlier, when she moved to Porter Ranch, she took regular long runs in the hills. Now she did not go outside except to walk Joey. In October, Joey began to vomit constantly, about thirty times a night. His gums bled. On the sidewalk, instead of sniffing the ground for unusual odors, Joey paused to sniff the air. Vicki often returned home from running an errand to find the entryway coated with bile. She was convinced that gas had been leaking from Aliso Canyon for years and blamed it for her general physical decline: the raspiness of her voice, problems with balance, inability to focus, loss of appetite, ringing in the ears, shortness of breath, extreme headaches, pressure behind the eyes, the failure of her brain to coordinate with her feet, and weakened immune system. When she traveled, the symptoms dissipated. "I'm very isolated," she said.

Paula Vasquez found the smell of gas so strong that she called 311, certain there was a leak inside her house. Every member of her family—her husband, their adult daughter, and their thirteen-year-old grandson—came down with bloody noses, blurred vision, and nasal congestion. She stopped opening windows and refused even to venture into her newly remodeled backyard, in which she had replaced the grass with artificial turf to conserve water use. She hadn't considered going into the pool, "because who knows what chemicals are falling into it?" Her neighbors didn't seem to share her anxiety; she had watched them pick lemons from their tree. She didn't say anything, but she felt concerned for them. She worried there was gas in the fruit.

Vasquez had a warm, cheerful manner; horror did not come naturally to her. But the leak had brought premonitions and dark visions. She had made her grandson snap photographs with her phone while she was driving home one afternoon on the Ronald Reagan Freeway. In the sky above Porter Ranch, a heavy funnel of clouds was lit neon orange, "like a big atomic cloud."

"Creepy, huh?" she said. "But I don't know anything about science."

✦

Vasquez, like Goode, Weiler, Vicki, and the Chows, did not know until the leak that a major oil field lay at the edge of her suburban neighborhood. Kyoko Hibino, a designer for an architectural firm, had lived in Porter Ranch seven years before realizing that an oil and gas facility stood across the street from her house. She was alerted by a notice she received, in 2014, announcing the drilling of twelve new oil wells. With her partner, a sound engineer named Matt Pakucko, Hibino formed Save Porter Ranch, an activist organization that at first amounted to little more than a page on Facebook. They struggled to rally support. Porter Ranch was an insular community, particularly in the gated areas, where people did not often see, let alone befriend, their neighbors. "It is a quiet neighborhood," Hibino said. "Very quiet. *Very* quiet."

Hibino had become concerned about the gas fumes that drifted down from the mountain. She had suffered bloody noses for years, and the couple began to suspect that there was an unusually high rate of illness in the community. Pakucko figured it wasn't all the fresh air and sunshine that was making everybody sick. They were about to launch a new air-monitoring campaign ("If You Smell Something, Say Something") when Save Porter Ranch's Facebook page and email account were inundated with questions about a sudden, overwhelming smell of gas. It was so strong that Pakucko was afraid to light a match. He called the fire department; inspectors confirmed that there was a leak and said that SoCal-Gas was trying to plug it. The act of going outside to speak with the officers was enough to make him dizzy.

Two days into the leak, Pakucko called the SoCalGas emergency hotline.

"If you smell gas," said an automated voice, "press one."

Pakucko pressed one. After a bar of Kenny Loggins' "Nobody but You," a representative picked up.

"May I have the nature of your emergency?"

"What's this smell of gas outside our house all night long?"

"I'm not sure, sir. May I have your address please?"

He gave it. "We're next to the SoCalGas storage facility."

There was a pause. "It looks like they were releasing gas into the air in that area."

"Releasing gas and making us sick."

"Natural gas is nontoxic, sir."

"Do you know why I'm calling you? Because on your website it says that if you smell gas, evacuate and call 911."

"Because there could be a gas leak. Not because it's toxic, sir."

"I'm getting sick and my girlfriend's getting sick and I got four cats throwing up."

"Okay, then you're having a reaction to it. But it's nontoxic."

"A reaction to a nontoxic gas . . . Why are they releasing gas?"

"It's just something that they do periodically."

"This is normal for them?"

"It's normal. They do it every now and then."

"There's no problem up there?"

"No, sir."

"You're a hundred percent sure of this."

"Yes. I am. Sir."

In the following days Hibino suffered heart palpitations, severe headaches, and a bloody nose that could not be stanched. When she went to the emergency room, she was charged twenty-two hundred dollars and given no prescription, except to flee Porter Ranch.

We are a show-me species, wired to look for visible evidence of invisible harm. The impulse can lead a person to blame global warming for a balmy afternoon in February or make a climate-change denialist find vindication in a snowstorm. The world's largest natural gas leak might have created the yellow dots—a residue of the petroleum-laced slurry used to plug the leak—but it had no known effects on clouds or lemons. The most dangerous threats to our species are precisely those that are most difficult to visualize: long term, slow to emerge, amorphous. Such threats include not only warming temperatures but mutating viruses and political corruption and yawning economic inequality, and they tend, at least in the short term, to be invisible, dimensionless, and pervasive, like death. Like natural gas.

While the yellow dots were coming out of the atmosphere and staining Charles Chow's Brougham, the planet was enduring the warmest January on record. It was the fourth consecutive month in which global temperatures had beat historical averages by more than one degree Celsius, another novelty. This news, when publicized at all, tended to be accompanied by NASA's map of the world, overlaid with orange and red splotches denoting temperature increases; otherwise there appeared stock photographs of sunbathers or melting icicles. Then came February, the warmest month in history. The threat to human civilization was advancing more rapidly than at any time in the last sixty-five million years but all that Americans could see were splotches and sunbathers and icicles.

All that the residents in Porter Ranch could see during those months of bottomless dread were empty streets and mysterious white vans. They were desperate for answers: Was the gas making them sick? How could they protect themselves? Who would be held responsible? The personal-injury lawyers were well prepared. They offered clarity, assurance, optimism. They could predict the future: a profitable future for the residents of Porter Ranch. At the weekly informational meetings held in local churches and hotels, lawyers answered questions from the community, often for several hours at a time, while circulating client forms.

Two days after their driveway conversation on Thunderbird Avenue, Rick Goode and the Chows attended a meeting hosted by R. Rex Parris at the Hilton in Woodland Hills ten miles south, just beyond the blast zone. R. Rex Parris belonged to a consortium of law firms that had filed the first class-action complaint against SoCalGas and its parent company, Sempra Energy, the nation's largest natural gas utility, on behalf of thousands of Porter Ranch residents. The group's press release anticipated that the leak would end up costing Sempra shareholders "well over $1 billion." About twenty residents sat around conference tables, helping themselves to the complimentary coffee, bagels, and doughnuts. A young lawyer, who seemed to have consumed a large quantity of the coffee, stood at the front of the room, delivering her sales pitch.

"Anything SoCal tells you," she said, "don't listen to it. Everything they say means *nothing*."

She advised the residents to keep daily journals. They were to note each occurrence of a physical symptom or strange smell. They were invited to list all expenses incurred by relocation or illness. Someone asked whether he could qualify as a plaintiff even if he lived ten miles from Aliso Canyon.

"Nothing's been established," said the lawyer. "I've heard between five and ten miles. But we don't have the data yet."

"They claim it started on October 23," said one older woman. "But in April, my dog, a boxer, died within two weeks. I know it was the gas."

"It was earlier than October 23," said the lawyer. "I just don't know when. We want it to be as early as possible, so we can get much more money for everyone."

The residents nodded in approval.

The lawyer explained that about thirty attorneys were assigned to the case. The firms would receive thirty percent of any payout. They predicted a settlement.

In a heavy Russian accent, a man exclaimed, "I escaped Chernobyl for this?"

The man, Igor Volochkov, who bore a passing resemblance to Gérard Depardieu, had moved from Kiev, sixty miles south of Chernobyl, to Los Angeles in 1986, when his wife was pregnant with their son. They had escaped the Soviet Union to save their lives and the lives of their children. The similarities between Chernobyl and Aliso Canyon unsettled him. "In both cases, they tried to hide information from the public. And just as you couldn't feel the radiation, or see it, it is the same here, with the gas." He shook his head. "Except things were better in Chernobyl. Nobody believed the government, so information went mouth to mouth. Here everyone believes in mass media. If they don't see it, it's not happening."

Volochkov knew something was wrong in October when his parrot, Bon, dropped dead. He bought a new parrot, Gosha, and a parakeet, Margosha. They died within a month. The same thing had happened in Kiev, when nuclear radiation from Chernobyl killed his parakeet, Petruschka. Volochkov said his son had asked why they moved from one Chernobyl to another Chernobyl.

"Our destiny," Volochkov told him, "is to fight against Chernobyls."

Nearly a month into the leak, the county's Department of Public Health ordered SoCalGas to pay for new housing for those affected by the gas odors. Nearly six thousand households, about half of Porter Ranch's population, accepted the offer, moving to hotels, apartments, and furnished homes in surrounding neighborhoods. Many of those who moved away returned frequently, even daily, to retrieve belongings they left behind, prune the rosemary, or sniff for gas.

There remained Porter Ranch residents who believed that the toxic airborne event was utterly benign, at least beyond its effect on real estate values. Jerry McCormack, who lived down Thunderbird from the Chows, had not been sick and had rarely detected an odor. He thought the whole panic was "a lot of foolishness. This is not Fukushima. Everyone is out to get the gas company. The hysteria is proportional to the number of lawyers coming to town." He conceded that his wife, who was recovering from cancer, could "smell it quite well" and was concerned. Her oncologist advised her to leave.

Adam and Mindi Grant, a couple in their mid-forties, lived a mile from the leak site. Their three children continued to play basketball and swim outside every day. Adam had smelled the gas only once. He laughed about it with his wife. Whenever someone got a bloody nose, he joked, "It's the gas!" He did not consider it significant that bloody noses occurred in Porter Ranch regularly enough to allow for a running joke.

Adam taught world history at a local high school. Mindi was an insurance lawyer. Some of her friends had real symptoms; others, she thought, were malingering. "They're setting up for a money grab. They think there's big money, deep pockets. But they're going to have trouble showing causation."

"Had the smell been horrific," said Adam, "we would have relocated. But because it's not affecting us as a family, I'm a little lackadaisical about it."

If the smell of gas makes one person dizzy while the neighbor next door can't smell a thing, is one of them lying? If a woman does not actually inhale gas but develops headaches and nausea anyway, does she suffer any less? Disaster psychiatrists call this phenomenon "somatization," a word that has replaced disfavored terms like "psychosomatic" and "hysteria." Man-made disasters can inflict severe psychological damage on those who are not physically harmed, damage that often manifests itself in the same way as exposure. Fear makes you sick. Inhaling toxic gas causes the same symptoms as the *fear* of inhaling toxic gas: headaches, dizziness, and nausea.

Residents of Porter Ranch had reason to be afraid. Nobody could tell them what they were breathing. Methane was gushing from the leak, of this they could be certain, but methane was not what they smelled. Methane is odorless. What they smelled were mercaptans: sulfur compounds that are released in animal feces. Mercaptans are added to natural gas pipelines to provide an olfactory alarm in case a leak occurs, the way banks insert exploding dye packs into money bags. Inhalation of mercaptans can cause headaches, dizziness, and nausea, but, like methane, they were not known to cause significant long-term health effects. The major health concern about the leak was that other, more toxic gases might also be escaping from the bowels of Aliso Canyon— gases dating from its previous life as an oil field.

Chief among these was benzene, a known carcinogen. Los Angeles air, among the most polluted in the nation, has a background concentration of benzene between 0.1 and 0.5 parts per billion. The World Health Organization has declared that "no safe level of exposure can be recommended." A

month into the Aliso Canyon leak, readings taken by SoCal-
Gas near its facility found benzene concentrations fluctuat-
ing wildly, between 0.3 parts per billion and a nightmarish
30.6; in Porter Ranch readings shot as high as 5.5 parts per
billion. Other toxic gases—toluene, xylene, hexane, and hy-
drogen sulfides—were also detected at elevated concentrations. Maddened by all this uncertainty, Michael Jerrett, the
director of UCLA's Center for Occupational and Environmental Health, began two months into the leak to conduct a
more rigorous survey. He installed nineteen monitors, which
resembled bird feeders, in backyards across Porter Ranch, to
take continuous readings of various pollutants. He placed a
twentieth monitor behind the evacuated Porter Ranch Community School, beside the school's vegetable garden, where
the eggplants and cherry tomatoes had rotted on the vine
and the soil was littered with plastic bags. What he found surprised him: the air in Porter Ranch had unusually low levels
of toxins. The air was cleaner than in much of Los Angeles,
thanks to the fact that Porter Ranch had largely been abandoned. If residents were inhaling benzene, it was from their
own tailpipes.

The composition of the gas was only the beginning of
what the residents of Porter Ranch did not understand about
the invisible fumes seeping from Aliso Canyon. They did not
know how far the gas was drifting or in what quantities. It
seemed that the smell was stronger the higher you went up
the mountain, and at dusk and dawn, but there was no data
to support this. One had also to consider the complicating
mystery of the erratic local wind patterns, which resemble
those of no other part of the Los Angeles basin and change

direction capriciously. Nor did anyone know what had even caused the leak in the first place. More than two-thirds of the wells were built before 1980, a fact that disturbed Mel Reiter, the editor of the *Valley Voice*, a monthly newspaper that was the only local business to profit from the leak: it couldn't print enough pages to satisfy the demand from law firms for full-page ads. If one well was leaking, asked Reiter, what were the odds that thirty more were, or would soon?

Regulations were in place but nobody knew who could enforce them. SoCalGas, as a private utility, did not fall under the regulatory oversight of any single agency. The South Coast Air Quality Management District was responsible for investigating air-quality complaints, but it shared jurisdiction with the California Energy Commission; the L.A. County Department of Public Health; the Air Resources Board; the Public Utilities Commission; the Division of Occupational Safety and Health; the Department of Conservation's Division of Oil, Gas, and Geothermal Resources; the Office of Environmental Health Hazard Assessment; the L.A. County Fire Department; the Governor's Office of Emergency Services; and the EPA. After two months, as panic overwhelmed Porter Ranch, the L.A. County Board of Supervisors called for the creation of yet another regulatory "structure" to oversee gas storage facilities.

For most residents, all this confusion added up to a single fact: an invisible gas was threatening their lives. "We don't know what methane is," said Sam Kustanovich, a Belarusian pawnbroker who had the misfortune of buying his house two months before the leak was detected. "Nobody knows. It could mean explosions. Me, I'm afraid of explosions."

Who wasn't? But some fires took longer than others to ignite.

$$\oplus$$

The global climate, even in drought-stricken Southern California, was not a pressing campaign issue. A cloud of poisonous gas that caused mass vomiting and nosebleeds in a wealthy, vote-rich community, however, was a candidate's dream, and the procession of scientists and lawyers ascending upon Porter Ranch was trailed by a caravan of politicians. None came out in favor of mass nosebleeds. Though the Twenty-Fifth Congressional District reaches only its purple pinkie toe into Porter Ranch, Bryan Caforio, a Democrat, made the leak a central issue of his candidacy. The Republican incumbent, Steve Knight, who received campaign donations from Sempra Energy, initially expressed sympathy for SoCalGas, as if it were the victim of a heinous assault committed by Aliso Canyon itself, before ultimately calling for a congressional hearing. Even Michael Antonovich, a Republican county supervisor who voted reliably against environmental regulation, loudly proclaimed his determination to hold SoCalGas responsible. As Henry Stern, a Democrat campaigning for state senate, put it, "We're all feeding on it in a weird way. How often are there climate disasters in suburbia?" At community meetings Stern was struck by the way that local residents, many of whom identified as conservatives, had begun to question the wisdom of relying on fossil fuels. "Climate change is not a real thing for most of these people," said Stern. "But you change your mind quick when your kids are puking."

The only politician who failed to use the gas leak for political gain was the one whom Californians might have expected to have been the most engaged: the governor, Jerry Brown, who had staked his legacy on his efforts to make California a global leader in the fight against climate change. His Office of Emergency Services oversaw the various state agencies responsible for responding to the leak, and Brown himself had sent a letter to the chief executive of SoCalGas, demanding accountability. But he did not visit Porter Ranch until eleven weeks later, when he declared a state of emergency, toured the SoCalGas facility, and met privately with members of the neighborhood council. This was nearly a month after he attended the United Nations climate talks in Paris, where he boasted of California's emissions-reduction plan, the most ambitious in North America. Matt Pakucko and the other local activists blamed Brown for prolonging their misery. "There he was in Paris," said Pakucko, "saying look how green California is, while ten years of green stuff is going into the air right now."

In Brown's defense, the director of his Office of Emergency Services said that his boss had better things to do than visit Porter Ranch: "Let's face it: We deal with so many emergencies out here. This is not Vermont; this is not Oklahoma. This is a nation-state." Such bravado did not impress residents of Porter Ranch, particularly after the revelation that Brown's younger sister, Kathleen Brown, was a paid board member of Sempra Energy, having received more than a million dollars in compensation. She had a nearly million-dollar stake in the Forestar Group, a real estate and natural-resources company, which was clearing chaparral for Hidden Creeks Estates, a gated community of about two hundred luxury homes and equestrian facilities, adjacent to Porter Ranch, on property abutting Sempra's. These sparse publicly available facts seemed to those living beneath the canyon the few visible traces of a conspiracy as noxious as a swirling underground sea of poisonous gas.

It took nearly four months for SoCalGas to cap the leak. Ambient methane levels dropped immediately; concentrations of air toxins reached "typical levels for the Los Angeles Basin," which could not be said to be a particularly enviable condition, or even an improvement, but suggested at least that the worst was over.

"When things get back to normal," Henry Stern had asked, "will there be any test that will ever make people satisfied?" There would not. Weeks after the leak was plugged, many residents, including Rick Goode and Igor Volochkov, said they still smelled odors and suffered bloody noses, headaches, malaise. "Maybe SS-25 is capped," said Kyoko Hibino. "But there is something seeping up from underground. The smell is strong. It is out there still."

Six months later, SoCalGas pleaded no contest to a criminal charge from Los Angeles County for failing to report the leak immediately; it agreed to a settlement of four million dollars, or the cost of three five-bedrooms in Porter Ranch. A year later, SoCalGas settled a lawsuit with the South Coast Air Quality Management District for $8.5 million. A year after that, in August 2018, it settled lawsuits brought by the California Air Resources Board, the California attorney general, the City of Los Angeles, and again L.A. County for a total of $119.5 million. In 2019, a consulting firm hired by the California Public Utilities Commission concluded a three-year investigation with the publication of a twenty-five-hundred-page report on the Aliso Canyon affair. The cause of the leak, it found, was a corroded pipe. The facility's wells were riddled with corrosion, casing failures, and leaks. By 1980, the safety system for SS-25 had deteriorated to such a point that the Tidewater Oil Company abandoned efforts to repair it. "System apparently bad," a report concluded at the time. In 1988, a memo recommended the inspection of twenty wells, including the one that burst in 2015. None were inspected. SoCalGas had routinely failed to investigate leaks. The program manager for the Public Utilities Commission who monitored the closure of SS-25 contracted a rare blood cancer called hairy cell leukemia; he sued SoCalGas for damages. Fourteen thousand residents filed another two hundred and fifty lawsuits.

It was uncertain whether the maladies suffered by the residents of Porter Ranch would ever be linked to the gas leak. It was certain that the atmosphere would suffer long-term consequences, but they would be as indecipherable as a plume of colorless gas leaked into a windswept canyon. After the leak was capped, the aviator-scientist Stephen Conley, asked to put Aliso Canyon into perspective, called it "a monster. It throws off L.A.'s emissions for the year. It's a significant percentage of California's annual carbon budget." He paused. "But it's about 0.002 percent of the global methane budget. It's not like next year will be warmer because of Aliso Canyon."

Conley turned out to be right. The next year was not warmer because of the leak. It was not warmer because of the

car trips that Porter Ranch residents made to their temporary rental homes, or the Aliso Canyon gas they used to cook
their dinners, or the energy required to heat their abandoned
swimming pools. The next year wasn't warmer because of the
two hundred thousand airplanes passing through Van Nuys
Airport. The next year wasn't even warmer because of the
roughly one hundred and forty billion cubic meters of natural gas that oil companies flared into the atmosphere. But the
next year was warmer.

Part II

SEASON OF DISBELIEF

4

FRANKENSTEIN IN THE LOWER NINTH

"We have snakes," said Mary Brock. "Long, thick snakes. King snakes, rattlesnakes."

Brock was walking Pee Wee, a small, high-strung West Highland terrier who darted into the brush at the slightest provocation—a sudden breeze, shifting gravel, a tour bus rumbling down Caffin Avenue several blocks east. Pee Wee had reason to be anxious. Brock was anxious. Most residents of the Lower Ninth Ward in New Orleans were anxious. "A lot of people in my little area died after Katrina," said Brock. "Because of too much stress." The most immediate source of stress that October morning were the stray Rottweilers. Brock had seen packs of them haunting the wildly overgrown lots, prowling for food. Pee Wee had seen them too.

"I know they used to be pets because they are beautiful animals." Brock corrected herself: "They *were* beautiful animals. When I first saw them, they were nice and clean—inside-the-house animals. But now they just look sad."

Six years after Hurricane Katrina, the Lower Ninth had

become a dumping ground for unwanted dogs and cats. People from all over the city took the Claiborne Avenue Bridge over the Industrial Canal, bounced along the fractured streets until they reached a suitably unsupervised block, and tossed the squealing animals out of the car. But it wasn't just pets. The neighborhood had become a dumping ground for all unwanted things. Contractors, rather than drive to the landfill in New Orleans East, swept trailers full of construction debris onto the street. Auto shops, rather than pay the tire-disposal fee (two dollars a tire), jettisoned tires by the dozen. They were joined by burned mounds of household garbage, cotton-candy-pink tufts of insulation foam, turquoise PVC pipes, sodden couches tumescing like sea sponges, and abandoned cars. Sometimes the cars contained bodies. Near the intersection of Choctaw and Law, two blocks from where Mary Brock walked Pee Wee, the police had discovered an incinerated corpse in a white Dodge Charger on an empty lot. Nobody knew how long the car had been stashed there; it was concealed from the closest house, half a block away, by twelve-foot grasses. That entire stretch of Choctaw Street was no longer visible, having been devoured by forest. Every housing plot on both sides of the street for two blocks, between Rocheblave and Law, had been vacated. Somewhere in the weeds a cross marked the spot where Brock's neighbor had drowned.

It was misleading to talk about abandoned lots in the context of the Lower Ninth Ward. Vast sections of the neighborhood

had been abandoned, so property lines tended to be indecipherable. (An exception was the sliver of land on the neighborhood's innermost edge, where Brad Pitt's Make It Right
Foundation had constructed seventy-six solar-paneled, pastel-
hued homes—though this seemed less a part of the neighborhood than a Special Economic Zone.) To visualize how
the Lower Ninth looked before the city's formal campaign
to clear the neighborhood, you had to understand that it no
longer resembled an urban or even suburban environment.
Where once there stood orderly rows of single-family homes
with driveways and front yards, there was forest. All of the
vegetation had sprouted since Katrina. Trees that did not exist
before the storm were thirty feet high.

The cartoonish pace of vegetation growth resembled
something out of a Chia Pet commercial but it was hardly
surprising to New Orleanians long accustomed to roads
warped by roots and yards blitzed by volunteers. The soil in
the Lower Ninth was extraordinarily fertile, thanks to centuries of alluvial deposits from the Mississippi River, which
formed the neighborhood's southern boundary. From the
river, the neighborhood descended like a ramp to the open-
water, brackish marsh of Bayou Bienvenue. The back section
of the neighborhood, which at its lowest point was four feet
below sea level, was the most devastated by the storm and remained the least inhabited, its population having decreased
by about seventy percent.

Many of the ruined buildings were cleared away in the
five years after Katrina, and most of the old foundations were
concealed beneath the overgrowth. The inhabited lots, about
one per city block, were the exception. With their dutifully

trimmed lawns, upright fences, and new construction, they stood out like teeth in a jack-o'-lantern. But wilderness encroached from all sides.

"You see rabbits," said Don Porter, who lived south of Claiborne Avenue, in one of the more densely occupied sections of the neighborhood. There were four houses on his block and only two were vacant. "You see egrets," he said. "Pelicans."

"A raccoon climbs on top of our roof," said Terry Jacko, standing with his younger brother Terrence in their front yard on Reynes Street. "It's huge. The first time I heard it, I thought it was a dude."

"I saw a possum in the backyard the other day," said Terrence. "Its teeth were about this big. I killed it with a stick. It was coming toward me, so I hit him. He just flipped over. I stayed inside after that."

There had been sightings of armadillos, coyotes, owls, hawks, falcons, and an alligator, drinking from a leaky fire hydrant like a child at a water fountain. Rats had been less of a problem because of the stray cats and the birds of prey. But it was not just animals that emerged from the weeds.

"Sometimes I see people coming out of there," said Terrence, pointing at the ruins of two houses, shrouded in weeds, across the street. "They're trying to get in my home."

Johnny Windsor, who lived in a rebuilt house on Lamanche and Rocheblave, surrounded on every side by forest, pointed out a thicket across the street: "They drag bodies in there." Bad things had been happening in the abandoned lots. While walking home from school one evening, a sixteen-year-old girl was dragged into a blighted house and raped. Windsor and his wife took turns sitting outside, keeping watch. "You never know," he said, "if someone's lying in the grass, ready to shoot."

Since Katrina, the neighborhood had undergone a reverse colonization—wilderness conquering civilization. Residents had fought back with hatchets and string trimmers to rebuff the settlers: southern cut-grass, giant ragweed, golden rain tree. But the effort had been futile. The lots required constant vigilance; vigilance required occupancy. A lot left untended for three months grew thick with knee-high weeds. After five months, saplings began to rise. A Chinese tallow tree, one of the neighborhood's more aggressive invaders, will grow from seed to two-foot-high sapling in a summer and within a year will be taller than the average man. By the storm's sixth anniversary, it had become clear that wilderness had triumphed.

The following month, the mayor of New Orleans, Mitch Landrieu, announced what amounted to a troop surge in the battle for the Lower Ninth. He called it the Nuisance Lot Maintenance Pilot Program. It was the city's third attempt to clear the overgrown lots in the Lower Ninth. The first contractor was dismissed after it was revealed he had served time for a felony. The second contractor was assigned to clear each lot only once; by the time he had gotten around to the final lots, the first ones had already reverted to wilderness. Now the city would handle the job itself. It hired a crew of twelve men—residents of the Lower Ninth or ex-offenders—to wage a block-by-block campaign to reclaim the neighborhood.

To understand why New Orleans had ceded an entire neighborhood to nature for six years, it was necessary to revisit a chapter from the post-Katrina era so painful that few residents had the stomach to discuss it. The Tulane geographer Richard Campanella called it the Great Footprint Debate. With most of New Orleans in ruin, the city had to decide how, and what, to rebuild. Which areas should receive priority, and which should be left for later, or never?

According to some citizens (predominantly white, affluent, power-broker types), the problem was mathematical. In 1960, the population of New Orleans peaked at 627,525. To accommodate the boom, the city expanded into low-lying marshland that was previously considered unfit for human habitation. Katrina hit the newer, lower-lying neighborhoods hardest. The Lower Ninth was not, contrary to national assumption, the city's lowest-lying neighborhood; it is higher, on average, than New Orleans East, Gentilly, Broadmoor, and Lakeview. But after the storm the poor, predominantly Black neighborhood could least afford to pay for its own recovery, was forced to contend with the greatest bureaucratic delays and the most fraudulent contractors, and received the least local and federal disaster aid.

In 2006, a year after Katrina, the city's population plunged to about two hundred thousand. Since 1970 the street grid had increased by more than ten percent. Could a city built for seven hundred thousand people maintain a population less than a third of that size? Could a shrunken taxpayer base afford to maintain miles of perpetually eroding streets or citywide services like garbage removal, policing, sewer pipes? If not, what should be done with the lowest-lying areas and their exiled residents?

A panel commissioned by Mayor Ray Nagin, Landrieu's predecessor, recommended converting large sections of the

hardest-hit neighborhoods into "green space." This activated the survival instinct of local community groups, who
saw it as a covert attempt to eradicate the city's poorest and
Blackest neighborhoods. These suspicions were not allayed
by statements like the one made by Joseph Canizaro, the
property developer who led Nagin's panel. "I don't want to
see people rebuilding on quicksand," he said, neglecting to
note that the designated "quicksand" neighborhoods housed
eighty percent of the city's Black population. "As a practical
matter, these poor folks don't have the resources to go back to
our city just like they didn't have the resources to get out of
our city. So we won't get all those folks back."

Ultimately the technocrats were demolished by public
sentiment in a city that even after the flood remained majority poor and majority Black. Nagin rejected his commission's
recommendations and adopted in their place an approach
that most generously might be described as laissez-faire. Residents were allowed to return to the Lower Ninth—which
was more than four times the size of the French Quarter—as
they desired, and the city's full footprint would be preserved.
But most residents hadn't returned—some because they
could not afford to return, others by choice. Seven hundred
people sold their land to the state. The neighborhood had no
police or fire station, hospital or supermarket. Between its
light population density and lack of basic services, much of
the Lower Ninth had fallen into the very condition that New
Orleanians after the storm were desperate to avoid: it had become green space.

Landrieu, who beat Nagin in 2010, had campaigned as
defiantly pro-Footprint. "To shrink the city's footprint," he
said, "is to shrink its destiny." After his election he earmarked

considerable, even disproportionate, amounts of federal and local financing to construction projects in the Lower Ninth: sixty million dollars for street repairs, fifty million for rebuilding schools, and nearly fifteen million for a new community center. But clearing the lots was the logical first step. To repair a street, the city had to be able to find it.

When Landrieu took office, New Orleans had the highest percentage of blighted properties of any American city—higher than Cleveland, Flint, and even Detroit, which, during the previous sixty years, had lost more than a million people. Unlike those of the Rust Belt cities, the population of New Orleans after the storm was growing. Landrieu emphasized blight reduction and won the public's adoration; in polls his favorability rating approached ninety percent. "The people of New Orleans have decided to rebuild every part of New Orleans," he said. That decision "necessarily put us on a longer course and created a challenge that is hard to achieve, especially with a limited amount of money. But I think that you can, with a lot of good strategy and thought, rebuild neighborhoods. We're in an exercise now that's attempting to prove that point." This exercise would begin with cutting the grass.

Richard Campanella, the Tulane geographer, operated on a different time frame. The walls of his office were decorated with giant photographs of New Orleans, the Mississippi River, and the United States taken from outer space. As a geographer, he had the luxury—or Cassandra's curse—to contemplate the city's future three decades, even three centuries from now. "The debate about the footprint is history," he said. "You can't reintroduce that question, given that the city, the state, and the nation as a whole has already committed recovery dollars to rebuilding houses and fixing utilities. To go back and reopen the wound—it's too late. The baby's already born. Maybe next time we could revisit this. I hope there isn't a next time. But of course there will be."

Few cities have been more acutely influenced by their geographic location than New Orleans, which the geographer Peirce Lewis famously called the "inevitable city on an impossible site." Here, location was destiny; or, more precisely, elevation was destiny. The oldest, wealthiest, whiter, "historic" neighborhoods, where most transplants resided, were on high ground. Most of the newer and poorer neighborhoods, where the greatest concentration of native New Orleanians resided, were at, or below, sea level—and sinking. The difference between the back of the Lower Ninth and the French Quarter, which survived Katrina relatively unscathed, was nine feet. Campanella's geographic history of New Orleans, *Bienville's Dilemma*, was an essential text for those new to town, and readers often asked him where they should live. He told them, "The closer to the river and the closer to the historic urban core, the less vulnerable you will be." Campanella, who was himself born in Brooklyn, possessed the transplant's unrestrained ardor for his adopted home; he wrote a newspaper column about eccentric episodes in New Orleans history. "The population we have now is roughly the population we had a hundred years ago. I would love to see our current population living on higher ground, in neighborhoods teeming with life—people on the streets, walking and taking the streetcar to work, living within proximity of most of their needs. I realize that we cannot go back in history. There are hundred-year-old cultural transformations that you can't just ignore. But that is what I would hope for."

Campanella was a beloved public figure in New Orleans, but such statements spoken by a city council member would be grounds for impeachment.

While Landrieu began his Marshall Plan for the Lower Ninth, the neighborhood's ruination began to attract researchers from around the world in the burgeoning field of catastrophe studies, a field with a bright—an incandescent—future.

Michael Blum, an ecologist at Tulane, called New Orleans "an outstanding arena in which to understand basic ecological principles related to disturbance." That was an understatement. The plight of the Lower Ninth could not reasonably be compared with the decline of neighborhoods in Rust Belt cities. The closest analogy to what happened in the Lower Ninth, said Blum, was a volcanic eruption on the order of Mount St. Helens. The next closest was the magnitude 9 earthquake that hit Japan's northeast coast in 2011 and sent a fourteen-meter tsunami over the Fukushima Daiichi plant, causing three nuclear meltdowns and the evacuation of one hundred and fifty thousand people. Katrina had not merely been destructive; it brought about a "catastrophic reimagining of the landscape." In much of the neighborhood, nothing remained—not people, animals, or plants. The ecological term for this was "simplification." "In 2007, before rebuilding started, it was like going to an agricultural field," said Blum. "Literally it was wiped clean."

What happened in the following years made the Lower Ninth one of the richest biological case studies in the world.

Ecologists had long hypothesized that after a catastrophic event human communities and ecosystems return at the same rate. But this theory had not been tested in real time. Blum was among a coalition of scientists—ornithologists, botanists, geographers, and sociologists—who had come to the Lower Ninth to learn how civilization will respond to the kind of ecological catastrophe that will dominate the future.

In the race between nature and human society, nature jumped out to an early lead. The pattern of growth was bizarre: the Lower Ninth was besieged by a feeding frenzy. A chaotic mishmash of species, many of which had never existed on that land or in proximity to each other, battled for dominance. Before the Lower Ninth was cleared for plantations in the mid-eighteenth century, the area divided into three ecosystems, correlating to elevation. The riverfront was lined with reeds and brambles; behind it stood a dense hardwood forest; and farthest back, where Mary Brock and Pee Wee lived, lay a cypress swamp interrupted by stands of palmetto. Very few species native to the land survived Katrina, mainly several kinds of sedge and aquatic grass and a handful of live oak, pine, and bald cypress trees. But opportunistic species intruded: crape myrtle, black willow, and golden rain trees laced with vines. The undergrowth sprouted with weeds as high as basketball hoops and flowering shrubs like lantana, oleander, and oxalis. Invasive species snuck in from the major avenues, the seeds transported by the flatbed trucks that entered the city from points east. The plant and animal life varied quixotically from plot to plot as the new species entrenched themselves, mustering strength, before fighting to annex additional territory.

This made the local ecosystem diverse but extraordinarily

unstable. It wouldn't last. "It's a very odd mix, one that you wouldn't otherwise see in nature," said Blum. "It's a Frankenstein community." Ecologically speaking, Katrina had created a monster.

The twelve men hired to tame this monster met in the Lower Ninth every morning at 7:30. They wore sunglasses, jeans, boots, and bright green city-issued T-shirts. On the back of each shirt was a fleur-de-lis; the front bore the slogan "Fight the Blight." When the crew arrived at a lot, several men tramped through the brambles, dragging to the curb any large pieces of garbage or tires they found. Then came the tractor, a two-wheel-drive Mahindra 4025, which a crew member drove through the property like a battering ram.

Enri Jacques, one of the older members of the crew, had not seen any bodies or skeletons, just rabbits, raccoons, and garter snakes. Jacques lost his home in the storm; he had to sleep on the Claiborne Bridge for four nights before being rescued by boat. He learned about the lot-clearing job from his probation officer. He had struggled to find work; six years after the storm his house remained unlivable. "This job," said Jacques, "is a blessing of Christ." He would cut until his city looked familiar again. He expected he might be cutting the rest of his life.

A property was crossed off the list once the grass was trim and the sidewalk clear of debris. A "cut" lot did not resemble in the slightest a mowed lawn. The fields were reduced to a pockmarked stubble, bald patches alternating with tussocks of jaggedly shaved weeds. In some lots the quantity of vegetation was so great that when the crew's work was done, the ground was overlaid by a heavy carpet of hay. The lots didn't last long in that condition, however. Plots cleared a couple of months earlier had already sprouted wildflowers a foot high.

After the tractor finished its circuit, two men strolled through the lot waving string trimmers. Adrian Tillman, a lanky, twenty-eight-year-old member of the crew, wrapped a black shirt around his head so that the flying debris wouldn't cut his face. Tillman had done construction work at Jackson Barracks, the military base in the Lower Ninth that was still undergoing repairs from Katrina. After being laid off, he struggled to find a job until his mother told him that the city was looking for Lower Ninth residents to cut grass. Tillman was well prepared; he constantly attacked the grass around his own house. His neighbors thanked him. "When they see us coming," said Tillman, "they clap their hands."

A man who called himself Mr. Harris was not one of them. He stood on a viewing platform built on the flood wall at the edge of Bayou Bienvenue. He ate sunflower seeds while three friends, lifelong residents of the Lower Ninth, baited lines for redfish. Mr. Harris gestured at a cleared lot near Florida Avenue. The crew had left behind a pile of construction debris.

"If we're going to pay money to do that, I want professional work," he said. "I don't want it to look like that there."

He spit his sunflower shells in disgust. A luxury motor coach, filled with tourists behind tinted windows, trundled down Florida toward the Make It Right houses. Sixteen expletives have been omitted from the following paragraph.

"Every day twenty tour buses come down this street to look at this neighborhood and take pictures," said Mr. Harris.

"Don't tell me they're touring the city. If you're trying to tour the city, then you're in the wrong neighborhood. They just ride around in the part that's been devastated. Lower Ninth Ward isn't receiving a single penny for that. Why can't I get something? I'm not a guinea pig. I don't want to be put under a microscope. We're the ones that suffered down here, who lost everything. There are still dead people that they haven't accounted for. It's frustrating. It took almost seven years for the Ninth Ward to look like what it looks like now, and it still don't look like shit."

The going rate for a Hurricane Katrina tour of the Lower Ninth Ward was forty dollars. Motor coaches were operated by Big Easy Tours, Historic New Orleans Tours, and Gray Line, which offered customers an "eyewitness account of the events surrounding the most devastating natural—and man-made—disaster on American soil!" Tauck-guided tours spent one morning in the Lower Ninth as part of its eight-day New Orleans package (from twenty-seven hundred dollars). The tour began at Greater Little Zion Baptist, a humble boxy church founded in 1900 and rebuilt a year after the storm. The tour's forty-two members, almost all white and older than sixty, sat patiently in the pews, cameras in laps, while images of destruction were projected on the sanctuary. An introductory talk about the horrors of Katrina—the unmarked graves, the toxic mold, the cruel bureaucratic inanities of FEMA—was given by Laura Paul, a Canadian who moved to New Orleans as a volunteer aid worker after Katrina. In her previous life, Paul had been a client coordinator for global express aircraft at Bombardier Aerospace outside Montreal. Now she ran lowernine.org, a nonprofit organization whose volunteers rebuilt homes, operated an urban farm, and collected the data on abandoned lots that were used to develop the Nuisance Lot Maintenance Pilot Program. "The issue is not the tours themselves," said Paul, "but the fact that people are making bank on them and not giving anything back to the community. They can be disrespectful: people get out of the buses, trample private property, and take pictures." But she approved of Tauck, which donated twenty-five dollars per tour participant to lowernine.org. "I like the people they bring to the neighborhood: upper-class, predominantly white people. They have a lot of money." A woman from a Tauck tour once sent Paul a check for five thousand dollars. "Getting money after a storm is like shooting fish in a barrel. But long-term recovery? People just don't want to know how long it takes. The truth is discouraging. The problem is that someday—and this is already starting—people will be like, 'Seriously? Enough with the Katrina stuff. Please, just stop.'"

The only person on the bus who seemed truly uncomfortable was the tour guide. Renee Whitecloud, who grew up in New Orleans, carried in her wallet a photograph of her flooded street in the Broadmoor neighborhood. By the time she returned home, after the floodwaters receded, the mold had climbed to the second floor. The mold was "psychedelic: green, orange, yellow, all different colors." She developed asthma and allergies and suffered frequent migraines. "This is a hard day for me," she said. "Every time I do this tour, I have to revisit the trauma." When the Katrina video played in the church, she stood outside, alone.

Because the motor coach was too large to negotiate the

cratered residential streets, it kept to the major thorough-fares, making a rectangle around the most ravaged section of the neighborhood. At the Industrial Canal it paused for the passengers to photograph the site of the levee breach. As the bus passed the Make It Right houses, a teenage boy ran to the curb. The driver—whose own house was still gut-ted from Katrina—pulled over. The door hissed open and the boy jumped on. The bus filled with the kind of silence that follows a popped balloon. The boy held a carton of homemade pralines.

"Three for ten dollars," he said. "Buy one for a good cause."

The forty-two tour members, beset by a sudden paralysis, avoided eye contact.

"Like to donate a dollar? Anyone?"

Silence.

"Going once, going twice . . . Sold! To the guy in the black jacket!"

The man in the black jacket flinched violently.

"No? Okay then. Going once, going twice . . . Sold! To this woman here in the front row!"

No one offered the boy any money. After another excruci-atingly long pause, the bus discharged him with a pneumatic sigh and scuttled back to the French Quarter.

In three months the Nuisance Lot Maintenance Pilot Pro-gram cleared more than twelve hundred lots. Victory seemed in sight: the transformation of the neighborhood was stark. Ruined houses still tilted like aged prizefighters on nearly every street; the roads were chassis-rattling slalom courses; and there were few people, apart from the tourists, in sight. But no longer were there full blocks of uninterrupted forest. The least-populated areas of the neighborhood were desolate but neat. It was uncertain what would happen next. If the properties weren't suddenly reclaimed or bought, they would need to be cleared again. And again. As Jeff Hebert, the executive director of the New Orleans Redevelopment Authority, said, "Cutting grass once isn't really a good option." But it was the best they had.

When asked about a long-term plan, Mayor Landrieu could offer no specific answers. "We don't know what the end looks like," he said. "We think we know what the process looks like. We want to get those lots back in the hands of private property owners so that they can take responsibility for them. Anything we can do to make them attractive to private investors, we want to do." It was a high bar, "attractive to private investors," and it seemed unlikely that trimmed yards would clear it. Landrieu agreed, but he was stuck. While the city had made significant progress in fighting blight in other poor neighborhoods, the Lower Ninth had, as he put it, "become a symbol of New Orleans' rebirth—whether that's justified or not." Nor was it enough to restore the conditions that existed before the storm; those conditions were dreadful, exhibiting all the worst manifestations of decades of weaponized institutional racism. "I keep telling people that we're not putting things back like it was," said Landrieu. "We're building the city we want to become."

In the end, the city surrendered. It shut down the Nuisance Lot Maintenance program after a year.

Peter Yaukey, an ornithologist at the University of Holy
Cross, had surveyed birdlife in the Lower Ninth since Hurricane Katrina. On his first visit, a month after the storm,
he was not allowed north of Claiborne Avenue because the
authorities were still searching for bodies. He was stunned
by the silence. There was a complete absence of birdsong. You
could hear a boom box from three blocks away.

Many of the species once common in the Lower Ninth,
like mourning doves and house sparrows, had not returned,
or only in small numbers. But over the course of several
years, as the lots grew fuzzy with tall grass and venturing
saplings, giving ample cover to thriving rodent populations,
Yaukey observed that something strange was happening.
Large predators, which fed on rats and mice, began to arrive in high numbers. So high, in fact, that there was soon
a much greater concentration of shrikes, falcons, and hawks
in the Lower Ninth than would be found in any rural environment. He suspected that barn owls were building nests in
the blighted buildings. The word Yaukey used to describe the
concentration of raptors was "supernatural." Both senses of
the word applied.

Yaukey began spotting birds through the windshield as
soon as he turned off Claiborne. He wore a pair of binoculars around his neck, but he didn't need them to identify the
species that glided hundreds of feet overhead: a white ibis, a
red-tailed hawk, a turkey vulture. At street level he pointed
out blue jays, cardinals, American crows, eastern phoebes,

killdeer, loggerhead shrikes, kestrel falcons, bronzed cowbirds, and, rarest of all, an open-ground woodpecker. A great egret, regal and stiff, promenaded down Choctaw, stalking lizards. A flock of three hundred European starlings pecked at seeds on the Caffin neutral ground. At the western end of Dorgenois, an exceedingly plump red-shouldered hawk perched on the bending branch of a mulberry tree. To Yaukey's astonishment, the hawk did not budge even when he came within ten feet. "It looks pretty tame," he said. "Almost to the point of being goofy."

Yaukey was surprised by how substantially the city's maintenance crew had altered the landscape. Before the troop surge, the aggressive, omnipresent Chinese tallow, whose maturing leaves had begun to turn yellow, crimson, and purple, had dominated. If it was left alone another five years, Yaukey worried it might overwhelm the neighborhood. With many of the lots having been reset, however, a wider variety of plants had begun to thrive in the understory. He was especially eager to find a dense lot that he could search for birds, like sparrows, that favored brush. He found a suitable spot at Jourdan and Law—three contiguous abandoned lots, just a few blocks behind Brad Pitt's houses and across the street from the Industrial Canal levee. It was at this exact spot, more than six years earlier, that four of the concrete slabs cracked, the soil gave way, and the inundation of the Lower Ninth began.

No sooner had the car come to a stop than Yaukey burst out of the door, rollicking into the brush. Flustered starlings began to hop and dart about. Yaukey made birdcalls. *Pish-pish-spish-spish-SPISH*, he said. *Wee! Weeweewee!* And another that sounded like a rotating lawn sprinkler.

Within seconds he was twenty feet into the plot, obscured to his neck by high weeds. The birds called back to him. "Field sparrow," he marveled, cocking his head. "Swamp sparrow."

Pish-pish-spish-spish-spish-SPISH.

"I've gone whole winters without seeing a field sparrow. They simply do not winter in residential New Orleans. So this habitat . . ." He trailed off. The excitement was high in his voice. He had the bounding energy of a child let out to recess. The growth was so dense that it was impossible to find secure footing. Every step cracked a branch. Thorns tugged on pant legs. Vines noosed around ankles like booby traps. He kicked a block of concrete, concealed beneath a mound of dirt—the foundation of the house that once stood there.

"Rattlebox!" said Yaukey, pointing to an invasive South American tree in the middle of a thicket. Clusters of dark brown pods dangled from its stalks like earrings. Only a sandy tuft of Yaukey's hair was still visible. His voice was hard to make out beneath the humming of the forest. "Orange-crowned warbler!" he shouted. "Ruby-crowned kinglet!"

Finally he was too deep in the woods. It was impossible to distinguish which calls were his and which the birds'. He maneuvered around a stand of fifteen-foot Chinese tallow trees, the green and crimson leaves waving mournfully in the wind, and then he was gone. The wilderness had swallowed him up.

5

CHICKENS WITHOUT THEIR
HEADS CUT OFF

When Henry Park trained as a meat cutter in central Illinois, his mentor spoke of a time when the local grocery stores kept live chickens in the basement. Most families didn't buy chickens then; they raised their own. On the rare occasion when a chickenless customer called, the grocer clomped downstairs, slaughtered an animal, picked it, gutted it, and hauled the carcass up to the meat counter to wrap. That was hard for Henry to believe, but not nearly as hard as believing that his own son, seventy-five years later, worked in a laboratory in San Francisco, creating chicken in test tubes.

In 1969, Henry was hired to stock groceries by a small IGA store in Rushville, a farming town in the middle of what locals affectionately called meat country. More cattle lived in Schuyler County than people. Hogs outnumbered cows ten to one. The closest city was Peoria, ninety minutes northeast, not that the residents of Rushville particularly cared to visit. Henry had not set out to be a meat cutter, and certainly not a "butcher"—a term that, even after five decades on the job, seemed a little grand. When the IGA's meat cutter quit, the boss asked whether Henry wanted the job. Henry refused, saying he had no interest in carving meat for a living. "It pays ten bucks a week more," said his boss. "I'm a meat cutter," said Henry.

He spent most of his time breaking beef carcasses. When a customer ordered a steak, he'd plop the hind quarter of a cow on the chopping block, crack the sirloin butt from the loin, and crack it from the round. The process took about an hour. His life changed in 1974 with the innovation of boxed beef: wholesale cuts of subprime beef, precut and shipped to market in vacuum-sealed cartons. The product did not taste as fresh as the beef Henry butchered, but he accepted the trade-off. It saved hours of grueling labor. Henry had learned that when it came to meat consumption, flavor wasn't the sole consideration. Price mattered, convenience mattered, availability mattered. He sold comfort.

By the time Henry bought the grocery store from its owner, in 1989, his two sons were teenagers. He told them that they would inherit the business one day. Nate, the younger son, began working forty-hour weeks at the age of fourteen. He organized the stockroom, mopped, and packed carry-out bags and took them to shoppers' cars. By fifteen he had earned enough trust to join his father behind the meat counter. It was a hallowed arena, the IGA meat counter, as central to Rushville's civic culture as the church or the Rotary Club. Nate learned how to make the best sausage in central Illinois and how to grind hamburger by hand, blending in the tallow fats and the trim. "Don't overwork it," Henry would admonish his son. "Don't get too hot." Henry could judge on sight the fat content of ground beef to a percentage point. Every weekend father and son watched Jacques Pépin's cooking show on PBS.

Nate was startled by the intimacy of the conversations he witnessed between Henry and his customers. Over the years Henry had learned their likes and dislikes, their competencies and blind spots. Some would knock on the door to the meat shop and say, "Hey, Henry, what am I eating for dinner tonight?" He felt a heightened sense of responsibility on special occasions; he liked helping people "make the holidays right." Customers wanted to cook food that tasted like home. Henry taught them how.

Nate took to small-town grocery work. He was, as his father put it, "one of those guys that never met a stranger." If every stranger was a friend, every customer was family. The Parks were never closer than when they worked together, running the family business. Working with his sons was the joy of Henry's life. "Those," he would say, decades later, "were the good times."

Though Nate was careful to avoid broaching the subject with his father, he never considered a career as a grocer or meat cutter. At Southern Illinois University he majored in journalism and dreamed of being a writer. After graduation he moved to Springfield and joined a radio station; he worked as an engineer and served as the straight guy on a Saturday morning show. Henry missed his son, but he had the sense that Nate was happy. Nate was not happy. He quit radio and joined a friend's pool installation business. He began to suspect that he was wasting his life. He decided to become a firefighter. He moved to Austin, tested for the department, and was awaiting his first assignment when his father told him about the grocery war in Rushville.

The IGA was in jeopardy. Henry had outlasted all four of the other grocery stores in town, but a large chain had announced plans to move in. Henry had a choice. He could compete against the behemoth, which would grossly undercut him, or he could accept their offer to run the larger store. It was an agonizing decision: Henry was an independent man. He felt trapped. Finally he capitulated—better to run the conglomerate's shop than to lose his own. Nate said he'd move back home to help. For a little while at least, father and son were back in business.

They were soon out of business. After a dispute over the terms of the contract, Henry was forced to sell his share in the store. Like that, he was out of work. The sudden blow felt to the Parks like a death in the family. Henry was inconsolable. He had lost more than the family business; he had lost his identity. He began to speak wistfully of his trade, like a professional athlete grieving a career-ending injury. "I just wish I could have a knife in my hand again," he told Nate, "and do what I do."

Nate couldn't leave him. *There has to be a way*, he thought, *there must*—and he found it. He'd help his father begin again, this time in their hometown of Beardstown, twelve miles from Rushville. They would open a butcher shop together. Nate proposed that they call it Henry's Market. Against Henry's strenuous objections, the name stuck. "You're the star, Dad," Nate told him. "You're the one they want." The people had missed him. Nate had missed him too.

Henry's Market opened in 2006 a couple blocks from the

Illinois River, across the street from the town square. Though it was advertised as a butcher shop and deli, Nate added organic produce, dairy, and beer on tap. The building's previous tenant had been a Mexican restaurant, and there was a small professional kitchen in the back, so Nate figured they might as well cook lunch, and as an afterthought he bought a chicken fryer. It then occurred to him that if they were going to serve fried chicken, they ought to offer some sides, so he made mashed potatoes, biscuits, and three daily soups. The soups were wildly underpriced, because they were made from scraps from Henry's butcher counter: the prime rib eye went into the chili, the bacon into the beans. Nate had never cooked professionally and found, with some surprise, that he was good at it. Before long Henry's was serving one hundred and fifty tables on Saturday nights. About five thousand people lived in Beardstown. In its first year Henry's Market sold fifty thousand pieces of fried chicken.

Nate's success hatched what he called "a weird power struggle" between father and son. As the restaurant's popularity grew, it swallowed up counter space and square footage. Meanwhile the customers lost interest in the butcher shop. They came for meals, not to buy groceries; the last thing one wanted, even in the heart of meat country, after gorging on fried chicken or hamburgers, was to buy more meat for the refrigerator. Did it make sense to order the same quantities of sirloin when much of it went unsold and had to be used, at a loss, in soup? Nate figured that had he never started cooking, the butcher shop would have succeeded. But they couldn't go back. Father and son agreed they had to "lean into the money."

Nate's menu grew increasingly ambitious. He had begun

watching YouTube videos posted by Moto, a new restaurant in Chicago that had popularized the term "molecular gastronomy." Meals at Moto began with the presentation of a tasting menu that, after reading, diners were encouraged to eat. Some nights the menu tasted like a panini, others a tortilla chip to be dipped into salsa or a cracker to be ground into a gazpacho. The table candle was poured over the clam bake. There followed edible packing peanuts; a photograph of a cow that tasted like filet mignon; and the "oil spill course," inspired by the Deepwater Horizon disaster, in which a chicken consommé, tinted Gulf of Mexico blue, was blackened by a pour of squid-ink noodles and crumpled edible paper that floated like toxic debris. Dessert was hot ice cream or orange sorbet in the form of ground beef nachos. The dominant cooking utensils were liquid nitrogen, balloons, syringes, class 4 CO_2 lasers, and a Canon i560 ink-jet printer containing flavor cartridges. The owner and head chef, Homaro Cantu, described his culinary technique with terms like "transmodification," "software and hardware," and "flavor tripping." Dinner at Moto lasted six hours.

The videos baffled Henry. "I don't see the food in the food."

"It's there," said Nate. "It's got to be. If it wasn't good, nobody would buy it."

Henry wasn't so certain what people in Chicago would buy.

In 2008, when the economy collapsed, Henry's Market was forced to close. Henry had to accept a job at his old IGA, working for a new boss. Nate vowed to stay in Beardstown and help his father start over again. Henry wouldn't let him.

"You know you're going to culinary school, don't you?" he said.

Nate did not know.

Henry insisted. "You have to go," he said, and Nate left Beardstown, and his father, for good.

⊕

Nate attended Le Cordon Bleu in Chicago instead of the more prestigious Culinary Institute of America in New York so that he could be close to Moto. After graduation, while sleeping in the spare bedroom of a married friend with a toddler and a newborn, he joined Moto as an unpaid *stagiaire*. He earned the trust of Homaro Cantu and his chef de cuisine, Chris Jones, on his third day at work, when Moto's septic pit exploded. Nate was surprised to find the restaurant's two senior chefs in the basement bathroom, trying to repair a sewage clog. "I have a little experience with plumbing," said Nate. "Can I help?" Cantu told him that if he fixed the septic pit, he'd serve him the GTM (Grand Tour Moto), the three-hundred-dollar, twenty-course tasting menu. Nate put his arm into a garbage bag, reached into the pit and, while Jones gagged, pulled out glutinous cloth napkins, seat folds, and spoons. Cantu offered him a full-time job and the GTM. "That meal," Nate later said, "changed my life."

He soon learned that the essential element of Moto's magic act, the disseverance of form from flavor, was not a gimmick—or not *only* a gimmick. It was a bid for a radical transformation of the American food industry. Nate had understood since working behind his father's meat counter that people were drawn to certain foods, and certain cuts of meat, by powerful emotional, psychological, and cultural yearnings. The dishes at Moto forced the diner to question those yearnings. This was salutary because American food culture was deeply destructive. It was unhealthy to eat, ecologically ruinous, and barbarically cruel to the animals and workers who were sacrificed to its mass production. Cantu believed that by shocking customers, or even making them laugh, you could help them to understand that the modern food industry was a relic of an insensibly brutal past that required upheaval.

Cantu spoke like the founder of a start-up. "I'm a product developer first and foremost," he would tell reporters. He served as a consultant to NASA and SpaceX, which wanted to print food and grow vegetables in spaceships. Moto's diners were not representative of the American population, or at least ninety-nine percent of it, but that was inconsequential; they were the test patients. The restaurant served as a workshop for experiments that Cantu believed could transform the world—experiments that he would demonstrate on TED talks, *Iron Chef*, the cable show he hosted (*Future Food*), and *The Today Show*, where he taught Katie Couric to ladle liquid nitrogen. Cantu converted Moto's basement into a laboratory, with digital scales, centrifuges, viscometers, Erlenmeyer flasks, and a glowing, wall-sized periodic table. The chefs carried on with the giddiness of mad scientists.

Moto invented the first "bloody" vegetarian burger, composed of beetroot and glycerides to supply fat content. It created recipes for weeds found sprouting from cracks in Chicago sidewalks and grew herbs in a vertical aeroponic garden in its windowless basement. And it made substantial use of the miracle berry *Synsepalum dulcificum*, harvested from an evergreen shrub native to tropical West Africa. The berry contained a protein—miraculin—that bound to the tongue's sweetness receptors and was activated by the acid present in sour foods. Miracle berries did not themselves taste sweet, but they made sour foods taste sweet. Nate learned how a miracle berry tablet could transform several drops of lemon dripped onto a mound of nonfat sour cream into rich cheesecake. "After you experience what miraculin can do," said Cantu, "there's just no good reason why we would continue using sugar." He thought that miracle berries could cure America of its sugar addiction. He wanted to prove that you could make delicious ice cream without sugar, grow fresh organic vegetables without farmland, and grill delicious hamburgers without killing cows. An experiment only worked, however, if the imitation was indistinguishable from the original—or an improvement on it.

After two years at Moto, Nate helped open Cantu's second restaurant, iNG (Imagining New Gastronomy), and later served as chef de cuisine at another Moto spin-off. Both failed. Nate was named chef at a restaurant opened by different owners, only to leave after four months. His menu, featuring dishes like "Cup-O-Ramen" and "Ranch Potsticker," was "highly complex," as one of the owners put it in a press statement; they had wanted "an approachable neighborhood establishment." It was a low time for Nate, and for the Moto clique. An investor sued Cantu, claiming the chef had embezzled Moto's funds to pay for his personal travel, unrelated business ventures, and the settlement of a sexual harassment suit with an employee. The investor sent his complaint to the press and gave denunciatory interviews. A month later Cantu hanged himself.

Cantu died in a warehouse he had planned to convert into an organic brewery. His initial idea had been to invent a beer that would intoxicate drinkers for no longer than twenty minutes, which he hoped could bring an end to drunk driving. After that scheme failed, he had turned to ales that tasted like green tea, an old-fashioned cocktail, and maple syrup. Nate despaired that a figure as beloved and successful as Cantu couldn't survive the pressures of the industry. If Cantu couldn't escape, who could?

Chris Jones could. At a bachelor's party for a Moto chef in Chicago, Nate reunited with the man who had hired him. After seven years alongside Cantu, an appearance on *Top Chef*, and on the verge of starting his own twenty-seat bistro, Jones had dropped out of the Chicago scene. He accepted an offer at a start-up with the resolutely anonymous name of Hampton Creek and moved his wife and toddler to San Francisco. The CEO, Josh Tetrick, described Hampton Creek, in a Cantu-esque phrase, as "a tech company that happens to be working with food."

At the time, the tech company was little more than Tetrick himself—a thirty-one-year-old graduate of the University of Michigan Law School who could barely use a microwave, let alone cook a meal. While sleeping on his ex-girlfriend's couch in Southern California, googling phrases like "food science" and "culinary innovation," he had come upon a TED talk given by Homaro Cantu. It was titled "Cooking as Alchemy." Tetrick had never heard of Moto but he was "blown away" by Cantu's application of scientific methods to high cuisine, by the images of watermelon "sushi," carbonated champagne grapes, and the Cuban sandwich that was served in an ashtray, disguised as a half-smoked Cohiba cigar. When the video cut to shots of Moto's basement laboratory, Tetrick paused and zoomed in, trying to make out, in the background, the names of the scientific instruments. He was determined to buy everything Moto had.

In those days Tetrick called his company Beyond Eggs. As his goal was to end the practice of killing animals for protein, he had decided to begin with the most ubiquitous form of animal protein in the world. He had only a mission statement—"To find a plant that scrambles like an egg"—but that was enough: it won him half a million dollars in venture capitalist funding. Money in hand, Tetrick set about trying to find somebody who could tell him how he might be able to achieve such a thing. After seeing the TED talk, he paid Cantu a retainer.

With the help of Chris Jones, Cantu prepared for Tetrick a series of long-shot ideas, which ended up going nowhere. Cantu was hampered by a maniacal fixation on designing square plant-based eggs, in the hope of making them easier to ship. "Hawks' eggs are square," Cantu told Tetrick, "so they don't roll down the hill." (Hawk eggs are ovoid and are laid exclusively in nests.) Tetrick decided it would be better to bring the experts in house.

He had the funding to do it, ultimately raising more than three hundred million dollars from Marc Benioff, Peter Thiel, Vinod Khosla, Jerry Yang, Eduardo Saverin, and Asia's richest person, Li Ka-shing. As many accelerating start-ups do, Tetrick hired executives away from Amazon, Apple, Google, and Netflix. But he also hired a battalion of professional chefs with Michelin stars, favoring those with expertise in molecular gastronomy. Jones was the first hired, in the fall of 2012. Tetrick assigned him to develop Hampton Creek's major debut: an eggless mayonnaise. By the end of 2013, Just Mayo could be found in the aisles of Walmart, Dollar Tree, and Costco. When Nate tried a jar of Just Mayo, he could tell that it was a Chris Jones production: brightly acidic, with a heavier viscosity than a typical mayonnaise, balanced, "crafted." It tasted like something concocted in Moto's lab.

Hampton Creek was, in the strict Silicon Valley tradition, a self-professed world-saving enterprise. Tetrick boasted of his fierce determination to fight animal suffering and the Grand Guignol horrors of industrial agriculture. The business was also, in the strict Silicon Valley tradition, plagued by errors of hubris, judgment, and technique. There were misguided forays into baked goods; investigations by the Department of Justice and the Securities and Exchange Commission (later dropped); a lawsuit from Unilever, the maker of Hellmann's, disputing the use of "mayo" for an eggless spread (also dropped); rumors of financial hemorrhage, unsafe work environments, scientific incompetence; comparisons to Theranos; and the mass departure of executives and the resignation of Tetrick's entire board. Still Hampton Creek continued to expand its enterprise. After mastering eggless mayonnaise, it turned to eggless eggs, and its valuation crested over one billion dollars.

If the eggless products were the appetizer, the main course was Project Jake, which was named, like Hampton Creek, after a dead dog. Hampton was a St. Bernard belonging to the co-founder Josh Balk; Jake was Tetrick's golden retriever, a regular presence at company headquarters who occasionally snagged food prototypes out of the lab. The goal of Project Jake was to enable the mass production of meat without the slaughter of animals. The intended consumer class was the high percentage of meat eaters who did not care about health or animal welfare but ate meat because of "the stories they tell themselves about it"—stories of virility, strength, tradition, nostalgia. Tetrick, who grew up in Birmingham, set himself the goal of making meat that would be consumed by "folks in rural Alabama." They would not eat plant-based burgers, no matter how well they approximated the real thing. But they would, Tetrick believed, buy burgers composed of cultured animal cells, if they tasted good enough. "For us to have the transition we need away from manufacturing animals and housing them the way we do," said Tetrick, "cultured meat needs to be more than an option on the menu. It has to be the *only* thing on the menu."

Cultured meat had not, by 2021, received regulatory permission to be sold in the United States—or any other country—but Tetrick, like the other two or three dozen companies racing to bring cultured meat to market, was betting several hundred million dollars on that changing soon. Tetrick expected it would first be legalized abroad, with Singapore the most likely debut market. Above the door to the meat laboratory, he had once put a sign: DESTINATION: WORLD'S LARGEST MEAT COMPANY BY 2030. He predicted there would come a day, between the years 2040 and 2050, when the majority of the world's meat would be produced without killing an animal.

Nate Park believed him. It was time, he thought, "to be part of a bigger movement." Why not apply the same level

of care and ingenuity that Moto dedicated to feeding a couple hundred people on a Saturday night to a couple million
people, or a couple billion? A few months after leaving his
restaurant, and a month after Cantu's suicide, Nate, who had
never been to California, told his father that he was moving
to San Francisco. He gave up being "Chef Nate." His new title
was "product developer."

Nate's first assignment was to improve the eggless egg. He
was impressed by the product's functional resemblance to
the real thing—it scrambled, it baked, it fried—but it didn't
taste great. He worked on Egg 2.0 in consultation with teams
of biochemists, molecular biologists, tissue and bioprocess
engineers, and food scientists. The Hampton Creek lab made
the Moto lab look like Mr. Wizard's World. There were cell
viability analyzers, rheometers, gas chromatographs, mass
spectrometers, and a plant library, overseen by the former
lead data scientist for Google Maps, that cataloged thousands
of samples obtained from more than sixty countries. The
plants were screened by robots, employing machine learning
and other forms of artificial intelligence, to identify novel
proteins. Nate found it thrilling, this sudden leap into the future, but on Saturday nights, back at his apartment, he found
himself suffering a form of withdrawal. He missed the buzz
of restaurant life: the anticipation of a big night, the quiet
tension before the eight o'clocks arrived, the exhilaration
of a frantic kitchen, the satisfaction of executing a series of
complex operations to sublime effect. The dining room of a

popular restaurant had a certain comfortable twinkling hum all its own. He found himself planning new dishes, even concepts for new restaurants. "You know what we *could* do," he'd begin, only to be stopped by his wife's eye-rolling. His jitteriness yielded, in time, to a kind of mourning.

During Nate's first five years in California, his father did not visit. On the phone Nate was not particularly forthcoming about his job. "So you're growing chicken in a lab?" Henry once asked.

"*I'm* not," said Nate, "but someone is. My job is to make it taste delicious. Unfortunately, I'm putting you out of work."

"Is that right?"

"Pretty soon we won't need butchers anymore."

Henry laughed. "Maybe it really is the future," he said, "if you don't need a butcher."

Nate worked hard to make his father's calling obsolete. Every day in the laboratory at Just, as Tetrick had rebranded the company, Nate ate chicken—or chicken prototypes. He grilled cultured chicken, poached it, breaded it, fried it. He molded it into the shape of burgers and breasts and meatballs. As he worked, and ate, he often found himself thinking about Beardstown. Would his former neighbors eat cultured chicken? Would it sell at Walmart? Would it frighten them?

Chicken nuggets, Tetrick had decided, were the easiest way to introduce a skeptical public to cultured meat, especially because a nugget's claim to being chicken was tenuous.

Most nuggets found on supermarket shelves were composed of less than fifty percent meat; the rest was fat, ground-up bone, blood vessels, and nerves. Nate experimented with various plant proteins but found that once he increased the nugget to seventy-four percent chicken meat, it no longer seemed like an approximation. "At seventy-four percent," said Nate, "Pinocchio feels like a real boy." The whole business, he decided, came down to perception. If a farmer in Beardstown cannot tell the difference between a Just nugget and any other nugget, Just will have succeeded.

There was the customer's perception of the food on the plate and the customer's perception of how that food got there. Most people did not know what factory farms looked like; most had never watched a slaughterhouse documentary or a PETA video. Just was betting that customers wouldn't worry about lab-cultured meat for the same reason: most of them would not imagine, let alone see, the cultured meat product sloshing around bioreactors in a laboratory in downtown San Francisco. In response to unease about the prospect of eating chicken from a test tube, Just could counter with images of the messy horrors of industrial chicken farming. But it would try to avoid the argument altogether. Though Just emphasized to investors its moral crusade and technological sophistication, its commercial success would depend on its ability to blend into the supermarket display case. Most people don't want to see how their food is made, any more than they want to know where their waste goes. The American experience of food consumption runs from the shopping aisle to the trash can.

Just's main advantage, Nate believed, was taste. The lab

could isolate the flavor characteristics of a naturally raised chicken and apply them to their product in the same way that Frito-Lay could incrementally adjust the "cool ranch" quality of a Dorito. Nate still spoke with pride of his culinary work, using terms like "craft" and "artisanship." But his approach had undergone a radical transformation. He no longer dedicated himself to thwarting his customers' assumptions but to fulfilling them. He had traveled, like many mid-career artists, from avant-garde to populism. It suited him better. "I consider myself a pretty average person," he said. "If I have a forte, it's that I know what the average person likes to eat." His goal was to make the consumption of cultured chicken a frictionless experience, designing the product to be indistinguishable from any other chicken sold at Walmart—only slightly more delicious.

Henry Park, who in his retirement picked up shifts at the Walmart Supercenter in Beardstown, was deeply proud of his son. He still pined for the days when he worked with his boys, but he decided that Nate, in his own way, had fulfilled the family legacy. "My whole career was feeding people and that's what he's doing," said Henry. "He's feeding them something else. But it's the same business."

Still he doubted that cultured meat would go over in central Illinois. "This is where they'll have their toughest sell, out here in the country. Old guys will probably resist it—old guys like myself." His main complaint was that Nate called it meat. "If it couldn't walk into the room on its own," said Henry, "it's not meat." He figured Just would have greater success among the younger families in town, on whom such distinctions would be lost. "They'll see it for what it is: the evolution of a protein. As each new generation comes along, it won't even be considered odd. Know what will be considered odd? That people actually raised their own chickens and killed them. Soon we'll get to the point where nobody will believe that ever happened."

6

ASPEN SAVES THE WORLD

The true test of a society is how it treats its most vulnerable members. Aspen's dogs are treated with a level of deference that would mortify most Americans. They shimmy down Hyman Avenue in cashmere sweaters and Swarovski crystal collars purchased at C.B. Paws, Aspen's pet boutique. Canine guests of the Little Nell, a five-star hotel, are greeted with a Puppy Jet Lag Kit, personalized brass identification tags, plush beds, and a dedicated staff of walkers and sitters on call twenty-four hours a day; the dogs are invited to join their owners at the patio bar, where they can order, from the gourmet Dog Menu, beef tenderloin with scrambled eggs and brown rice. Aspen's dogs even have their own private jets. Not long ago a Gulfstream V jet landed at Aspen/Pitkin County Airport and a poodle was escorted down the gangway. The dog had been forgotten at its Upper East Side penthouse by its owner, who had sent the plane back to retrieve it.

Such profligacy is wasted not only on Aspen's pets. Most of Aspen's homes are second homes, or third or fourth or fifth homes, and it is common for them to be inhabited no more than two weeks a year. Yet they are programmed to run in perpetuity, like the automated house in Ray Bradbury's "There Will Come Soft Rains" ("The house was an altar with ten thousand attendants . . . But the gods had gone away, and the ritual of the religion continued senselessly, uselessly"). The heating runs in the winter, lest the pipes freeze, and the air-conditioning in the summer, lest the oil paintings drip. Based on energy usage, it is impossible for Aspen Electric to determine whether most homes are occupied. It is also impossible for potential burglars to determine, because, as a security precaution, homeowners keep their exterior and interior lights on year-round to simulate occupancy. In 2019 there were ten burglaries in Aspen. Police were summoned by alarm systems six hundred and twenty-nine times.

Aspen's homes—owned by Kochs, Bezoses, Waltons, DeVoses, Pritzkers, Lauders, and a commissariat of oil magnates—have solved problems that most Americans don't realize they have. To avoid the unsavory sensation of stepping out of a hot shower onto cold marble, the bathroom floors are heated; towel racks are warmed to the temperature of the shower water. Driveways are undergirded by snow-melt systems. The electric heating tape that zigzags around most Aspen roofs, which owners often neglect to shut off in the summer, alone can cost more than a thousand dollars in annual utility bills—particularly as the homes that the roofs cover are among the largest in the world. "Many of them are actually very efficient emission envelopes," said Chris Menges, the sustainability programs administrator in the city's Climate Action Office, "but when you're talking about five-thousand-square-foot-plus homes and hot tubs and electronics and movie-theater-style TV rooms . . ." Most local residences, if they are common enough to be listed by Sotheby's, follow adjectives like "premier," "quintessential," and "spectacular," as in "The Spectacular Compound in Wildcat Ranch," an eight-bedroom, sixteen-bath, twenty-six-thousand-square-foot single-family home, on sale for fifty million dollars. The average property on Red Mountain, which *Forbes* calls "Billionaire Mountain," goes for $2225 a square foot. But there are not many average homes in Aspen.

This tally doesn't take into account the ubiquitous SUVs, the cost of shipping goods over mountain to a remote valley (gas costs fifty percent more at the pump than in Denver), or the Gulfstreams and Bombardiers gilding the airspace. In 2019, Aspen/Pitkin County Airport welcomed more than thirteen thousand private planes. Aspen's population is 7401, not including dogs.

"The winter lasts forever," wrote James Salter in a 1981 ode to the town where he spent the last forty-five winters of his life. "Skiing begins in late November and continues until mid-April."

No more. About twenty years after Salter wrote those lines, Chris Davenport, a world-champion mountain skier, was hiking to Maroon Bells, Aspen's most celebrated landmark. Near the end of that hike, as you reach an altitude of fourteen thousand feet, Aspen's second-most-iconic landmark comes into view: the massive snowfield that gave Snowmass Mountain its name. Long considered the largest permanent snowfield in Colorado, Snowmass descended the mountain's eastern slope like a billowy white apron. It was a local tradition to hike the slope in July or August, have a picnic near the summit of North Maroon Peak, and ski down. But when Davenport, two hundred feet from the summit, looked for the familiar view, he became disoriented. Snowmass was missing. Had it moved? No, he realized, the mountain was still there. Only the snow was gone.

"I speak the language of the mountain," said Davenport. "She doesn't hide anything. When she speaks, it's not subtle."

Aspen is running out of snow. The ski season is more than a month shorter than it was fifty years ago. It is expected to be nearly two weeks shorter still by 2030 and will continue to shrink in subsequent decades. In the spring, wet slab avalanches—in which an entire slope slides down a mountain, devouring the trees, rocks, and ski lifts in its path—will become increasingly common. The slope most prone to avalanches lies directly above Spar Gulch, an intermediate ski trail that is one of Aspen Mountain's main thoroughfares, and one of the only routes off the mountain. The beginner slopes will become pockmarked with rocks and turf, requiring snow machines to make up the difference. By 2100, there will no longer be any natural snow at the base of the mountain. In high-emissions scenarios, there will be no skiing in Aspen at all, and the local climate will resemble that of Amarillo, Texas. Amarillo by that point will resemble Venus.

By then Aspenites will have greater concerns than ski conditions, or heat. Even in the worst-case scenarios, Aspen and the other mountain towns in the American West, despite warming more rapidly than global averages, will still be more habitable than cities in the plains. When Denver turns into desiccated, sunbaked desert, Aspen will seem refreshing by comparison. The problem is not the heat. It is the water. Aspen's streams and creeks are fed largely by snowmelt. Snowpack melts gradually, working as a natural pipette, supplying the streams with a steady trickle. As the winter shortens and the snowpack shrinks, the remaining snow will melt even more quickly. Streams will flow higher in midwinter, and they will run dangerously low in the summer, threatening the survival of the riparian ecosystem: its brown and cutthroat trout, chorus frogs, aquatic snails, and backswimmers; the common muskrats, American beavers, and moose that drink from the streams; the olive-sided flycatchers, Brewer's sparrows, and bald eagles that roost on its banks; and the wetlands and forests it nourishes. Eighty percent of Colorado's wildlife needs access to a riparian habitat to survive, but today these habitats make up only one percent of the state's land. Aspenites also need access to waterways to survive: two frail mountain creeks, Castle and Maroon, provide some of the city's energy and all of its drinking water. The Roaring Fork River, the repository of all of the mountains' streams, could evaporate entirely most years.

That will make it more difficult to fight the forest fires. The worsening droughts have already caused the cancellation of the July Fourth fireworks nearly every year for a decade. In 2018, Aspen announced it would replace the fireworks with a drone light show, only to cancel the performance after the sky darkened with wildfire smoke. Come the second half of the century, the fires will be smaller but will occur more frequently. Their destructive force will be amplified by increased outbreaks of insect infestation. As fewer winter nights drop below freezing, the populations of gypsy moths and bark-eating beetles will thrive. A ravenous population of pine beetles has been busily preparing the lodgepole pines that surround Aspen for future conflagrations.

Many of the trees are not waiting around for that eventuality. Species that thrived in cold temperatures and high altitudes, like firs and spruces, have begun migrating up the mountains, fleeing the increasingly inhospitable valley. The dominant vegetation type will shift from taiga-tundra to boreal conifer forest. Aspen is even losing its aspens.

This makes it all the more surprising that Aspen has taken it upon itself to save the world from environmental collapse.

"I get annihilated," said Auden Schendler, an executive at the Aspen Skiing Company, when asked whether he appreciates the irony. "People call me a hypocrite. They say, 'Shut up, you're from Aspen.' But who ought to lead, if not Aspen? Bangladesh?"

Schendler stood at the foot of a coal-powered gondola that, though it was the middle of summer, was running circuits up Aspen Mountain. He had been employed for fourteen years at Skico, as it was locally known, but Schendler didn't think of himself as a company guy. He preferred "dirtbag." Despite having been raised in Weehawken, New Jersey—or because he was raised in Weehawken—he fell in love with the western ideals of wild beauty and virtuous toil when he was fourteen, after his uncle took him on a punishing three-day hike through Montana's Bob Marshall Wilderness. Schendler's appearance struck a compromise between his nature and his nurture. As a concession to Skico, he was freshly shaved, with a crisp haircut and a pressed oxford shirt. But the shirt was tucked baggily into his jeans, and the sandals and black Oakleys belonged to the kind of person who sneaks away from his desk to take a twenty-five-mile bicycle ride in the mountains—as Schendler did whenever the opportunity presented itself, with his boss's encouragement. Like most Aspenites, Schendler was exuberantly, ostentatiously healthful: tall and hale, with a strong jaw, strong hands, and a strong brow. At forty-three he looked ten years younger and had the energy of a man twenty years younger. Auden Schendler was the skier or biker or marathoner who raced past you with a smile and a friendly wave, going twice your speed.

Schendler was Skico's vice president of sustainability, hired to reduce the company's environmental damage. This was a considerable responsibility, because Skico was the largest employer in not only Aspen but the entire Roaring Fork valley—the broad green plain that runs forty miles southeast from Glenwood Springs to Aspen, circumscribed, like arches on a crown, by fourteen-thousand-foot peaks. Skico operated all four local ski resorts, seventeen restaurants, and two luxury hotels. On Schendler's first day of work, he told the Little Nell's manager that he was going to replace all the lights in the hotel with compact fluorescent bulbs. The new bulbs would last ten times longer, save money, and cut energy use by seventy-five percent. It seemed an obvious baby step. The hotel manager flatly refused. "When you go to Las Vegas and stay in a Motel 6," he told Schendler, "it has compact fluorescent bulbs. This isn't a Motel 6."

Schendler concluded he'd have to take a more cunning approach, focusing on projects with less public visibility. Under his leadership, Skico built its own solar array, hydroelectric plant, and methane-capturing facility. After a decade, Skico's revenue grew by more than forty percent, while its emissions fell by about four percent (a significant reduction given that, thanks to reduced snowfall, Skico's snowmaking operations produced more than six times as much carbon emissions in 2016 as it did in 2000). The Little Nell at last even switched out its incandescent lightbulbs. Schendler wrote about his successes and blunders in *Getting Green Done*, a complimentary copy of which appeared in every room of Skico's Limelight Hotel. The hotel did not keep records of how many guests took the book home with them.

Having eliminated Skico's carbon footprint, Schendler established a new objective: "To make sure that Aspen Skiing Company stays in business forever." That was another way of saying that his job was to stop global warming. Schendler became the only corporate executive in the United States whose success was measured not in profit but in snowfall. His tone shifted accordingly. He became less accommodating. He was, as Chris Davenport put it, "ready to pick a fight any minute." One of Schendler's most public fights was with Skico's chief rival, Vail Resorts, which did not consider it shrewd business practice to discuss the consequences of a heating planet. After a study, funded by a nonprofit foundation that Schendler oversaw, calculated that warmer winters had cost the ski resort industry more than a billion dollars in the previous decade, Vail's CEO, Rob Katz, responded with an editorial in *The Denver Post*. "It's hard to understand how the weather changes the way it does," he wrote. "You can count me out of the group that says we need to address climate change to save skiing."

"They think we shouldn't talk about this," said Schendler,

"but it's not a negative message. We're going to save this industry."

He made more than a dozen visits to congressmen in D.C., gave lectures at the Googleplex and the Yale School of Management, and published essays in *The Atlantic* and *Grist* with titles like "Selma, Montgomery, and Climate Change" and "*The Wall Street Journal*'s Willful Climate Lies." He reserved his greatest scorn for the cosmetic, ostentatious gestures that corporations made to burnish their credibility as environmental stewards—the option to decline fresh linens—while failing to support serious reforms. The bluntness of his arguments damaged his relationships with colleagues, including his mentor, the scientist and environmental activist Amory Lovins. Before joining Skico, Schendler worked for three years at Lovins' Rocky Mountain Institute, which promoted the doctrine of "natural capitalism"—the belief that companies, and by extension the world, will inevitably convert to renewable sources of energy and conservation practices because, as technology advances, such behavior will become profitable. The argument might hold for lightbulbs, but, as Schendler wrote in one jeremiad, it failed to "deal with the problem at anywhere near sufficient scale." That essay was titled "Corporate Sustainability Is Not Sustainable." Lovins stopped returning Schendler's calls.

Schendler called for "a civil rights–style revolution in climate." In his analogy, Aspen would be Selma. "Aspen can tell a story. We have the money and access to the most influential people in the world—which is another way of saying the wealthiest people in the world. We have an opportunity to use Aspen as a weapon on climate policy. As a baseball bat." In this analogy, Schendler would be not a slugger but

8

a pesky leadoff hitter—say, the Colorado Rockies' Charlie Blackmon.

The premise of Aspen as a climate model for the world—a shining city at the bottom of a hill—was introduced in 2005 with the Canary Initiative, a series of ambitious resolutions to reduce the city's carbon footprint. A decade later, Aspen Electric completed its full conversion to renewable energy. Aspen customers pay one of the lowest utility rates in Colorado. Water conservation has been fortified by a retrofitting of the city's pipes. The high-tech bus system, free to ride within fifteen miles of Aspen, has reduced automobile traffic to the level of twenty years earlier. As the expression goes, it's the rich who have money. It's the rich who have cheap renewable energy.

Aspen was rich enough to make these concessions before it was cost-effective to do so. "We have the liberty to think about the big picture," said Mick Ireland, a three-time mayor who looked like Ronnie Wood, had the guitarist committed to a rigorous plant-based diet and five hours of mountain biking a day. "A lot of communities struggle with things like budget cuts, crime, and school shootings. Here, our basic needs are met, so we're able to take action on larger issues. We're obliged to lead by example because we can. We have a stage. If this were Carbondale"—a city thirty miles up valley—"Bill Clinton wouldn't be asking me to lunch."

This sense of enlightened sacrifice, of noblesse oblige, can be traced further back than the Canary Initiative, to the spring of 1945, when Walter Paepcke, a cardboard-box magnate from Chicago, arrived in town. There were no paved roads or stoplights, and the population had dwindled to seven hundred, but Paepcke found a dilapidated grand hotel, the charred remains of an opera house, brick-and-sandstone country stores, and Victorian mansions—the legacy of a silver mine that had been one of the richest in the world, with a forty-mile-wide mother vein. In 1892 alone, the equivalent, in today's dollars, of three billion dollars' worth of silver was extracted from Aspen's mountains. The Sherman Act was repealed a year later. Silver was demonetized and its value crashed. Nearly all fifteen thousand residents of Aspen fled, abandoning the city's treasures. Most valuable of these was a hydroelectric plant built to provide electricity to the mining tunnels. From a small reservoir three hundred feet above town, water descended a flume—a waterslide made out of wooden planks—into five turbines. Aspen was one of the first cities west of the Mississippi to be fully electrified. Incandescent streetlamps were erected and electric light twinkled through every window. The city was powered entirely by water until 1958.

Paepcke, a Yale graduate who incorporated modern art into his company's advertising campaigns and audited the Great Books seminar at the University of Chicago, dreamed of re-creating Aspen in his own image. After incorporating the Aspen Skiing Company in 1946, he grew more ambitious. "He saw it first as a ghost town worth preserving as such," said Robert Maynard Hutchins, a president of the University of Chicago. "He then began to think of it as an American Salzburg." In 1949, Paepcke invited artists, scholars, businessmen, and politicians to Aspen for a festival in honor of Goethe's two hundredth birthday. The Goethe bicentennial served as a model for the Aspen Institute, which Paepcke founded the same year, followed by the Aspen Music Festival and the International Design Conference. It wasn't so much a city that was emerging as a grand humanistic experiment—what came to be known as the Aspen Idea, defined as a "harmony between mind, body, and spirit." By 1953, two bookstores had opened downtown, along with boutiques selling Venini glass, Pucci dresses, Gucci leather, and Andy Warhol drawings. Paepcke's rehabilitation of "a shattered town into a kind of national treasure of arts and ideas," wrote the journalist Peggy Clifford in her 1980 memoir, *To Aspen and Back*, "was not an act of philanthropy. It was an act of passion—and hubris."

"Aspen has always attracted people who think they can do whatever they want to do," wrote Hunter S. Thompson, who first visited this "experimental behavioral tank" in 1960 and later settled fifteen miles away. "Of course," he added, "you can't create a valley for the rich and then expect to live in peace with them. The rich are monsters."

Thompson's spirit could be heard cackling in the heart of Aspen's downtown during a recent iteration of the annual Ideas Festival. The evidence of Aspen's enlightened environmentalism was visible everywhere: the ubiquitous recycling bins; the stands offering complimentary biodegradable doggie bags; the bottle refilling stations with signs nudging passersby to drink Aspen tap; the bicycles and cars rentable by the minute. Maria Shriver glided by with two friends. She looked happy. Everybody in Aspen looked happy. The Aspen Idea was thriving. The private jets lowering overhead contained many of the world's most prominent intellectuals, CEOs, and politicians. They would give talks with titles such as "Fear and Hope: Climate Change and Policy Solutions" and "What Is the Right Energy Mix?" By all appearances, Walter Paepcke's dream had been realized.

There was only one problem. No matter how effectively the town cuts its carbon emissions or preserves its pristine surroundings, Aspen, as we know it, is doomed.

The people who best understood what has happened to Aspen, and what will happen to the American West in the decades ahead, lived just over the mountains, in a former ghost town called Gothic.

You can drive to Gothic from Aspen, but it is nearly as fast to walk. If you hike to Maroon Bells and keep going, over West Maroon Pass and down the other side, you descend, several hours later, into East River Valley. Widening and deepening as it goes, the valley follows its river four miles to the Crested Butte Mountain Resort, and in another four miles spills into the town of Crested Butte. Gothic is hidden at the valley's highest point, nearly ten thousand feet above sea level and invisible to civilization. It lies near the end of a dirt road that is impassable for six months a year. Gothic's permanent population was one.

That person was Billy Barr, who moved there in 1973. Barr looked exactly how you might imagine a man to look had he spent forty years in a cabin in the Rockies, surrounded by mountains and snow. He looked like mountains and snow: craggy features, with stringy white hair that hung to his shoulders and an exaggerated white beard. He inhabited the most remote of Gothic's several dozen wooden cabins, some dating to the nineteenth century, that climbed the hills on either side of the dirt road and recessed into the forest. At the end of autumn, like the other mammals in the surrounding hills, Barr stockpiled enough food to last the winter. When he desired a vacation, he skied four miles to the paved road, waited for a bus to Crested Butte, and switched to another bus to Gunnison, where he spent the night. He didn't leave Gothic often. "Everyone has this idea: sitting in a comfortable chair in your cabin, reading a book, with the snow falling softly outside. The truth is, it's boring as shit. But I like it."

Each day for more than forty years Barr had made entries in a weather journal. He noted the time of sunrise (he rose long before the sun) and sunset. He wrote observations about the quality of light, cloud cover, and wind strength, using a numerical rating system of his own invention. He calculated snow depth with a pole; he measured snowfall on a snowboard that he cleared twice a day; he assessed snow density with a hanging butcher's scale. He marked the first appearance of the local fauna in the spring—the ground squirrel, chipmunk, robin, and red-shafted flicker—and their final appearance in the fall. In a separate journal he kept detailed accounts of avalanches, about four hundred of which he observed each winter. Taken together, his journals described the radical transformation of Colorado's high alpine landscape in the previous half century. Barr had managed, accidentally, to create one of the most comprehensive climate databases in the world.

When the snow melted off the dirt road, Gothic was transformed into the Rocky Mountain Biological Laboratory, or RMBL (pronounced "rumble"), one of the continent's leading research field stations—"geek camp for scientists," as one camper put it. Joining Barr, who served as RMBL's accountant, were some forty scientists and forty graduate students, as well as visiting groups of students of all ages. Although the scientists came from many disciplines—particularly botany, ecology, and evolutionary biology—they inevitably found themselves talking climatology.

The pollination biologist David Inouye had for forty-three consecutive summers come to Gothic to observe its wildflowers. His goal, when he began the experiment as a graduate student, was to understand how flower populations changed over time. He returned each summer to the same meadow of lupine, American vetch, and bluegrass. Each day he counted every flower. As the summer crept earlier, the flowers bloomed sooner. The trigger, Inouye figured out, was the date of snowmelt. He used to be able to travel to Gothic from the University of Maryland after spring semester ended. Now he had to send a flower-counting assistant weeks ahead of him, for the melt occurred while school was still in session.

An even older study had tracked Gothic's marmots, which since 1962 have lived in a surveillance state. Each marmot, at some point in its life, had been trapped and marked with a symbol drawn by a toothbrush dipped in black dye. Graduate students hiked daily into the hills to spend hours staring at the critters through binoculars. The students took busy notes about the behavior of the animals, which were named after their marks: Smiley Face rubbed its cheek on rock. Musical Note pushed Mickey Mouse off rock. Lollipop mounted Sail Boat. There were sixty years of notes like this. We tend to think of evolution as a process that occurs over millennia, but after more than fifty years natural selection can begin to reveal itself. And the marmots were changing. They were getting fatter. In forty years, their hibernation time decreased by forty days—one day less per year. With longer time aboveground, the marmots ate more, which made them more likely to survive the hibernation and also more likely to be nabbed by a coyote.

The world's longest-running experiment on the effects of global warming had unfolded several hundred yards above RMBL's cabins, where heat lamps dangled from a web of cables over a gentle slope. The lamps, which had shone continuously since 1990, warmed the ground by two degrees Celsius, the minimum amount by which the planet's temperatures will increase this century. The lamps played a trick on the wildflowers. In early spring, they caused the flowers to bloom prematurely, only to wither before the summer rains came. Some species had dwindled, and hardier ones, like sagebrush, had annexed territory, but overall there were fewer flowers. Crested Butte, the state's Wildflower Capital, was losing its wildflowers. Colorado was losing its forests. And Aspen was losing its snow.

It was Auden Schendler's conviction that for Aspen to inspire a global energy revolution, it must not only tout its successes but acknowledge its failures. He was well prepared for failure. What he wasn't prepared for was the *way* in which Aspen failed—a failure so spectacular that it threatened to undo all the good work of the past decades. In their race to convert to renewable energy, Aspen's environmentalists had divided into camps and began fighting each other. "We were eating ourselves alive," said Schendler. The internecine crisis became so bitter that in Washington political circles the Aspen Idea was replaced by a new term: the Aspen Problem.

Like all western problems, Aspen's came down to water. For Aspen to rid itself of coal-fired energy, the city determined that it needed to build a new hydroelectric plant. There was an obvious location: the bank of Castle Creek, where the original plant had stood for more than a century. Schendler was one of the plan's most aggressive boosters. The city hired ecological consultants to study the proposal, write environmental impact statements, and develop regulations to ensure that the plant would not harm the health of the creek. In a 2007 referendum, more than seventy percent of the electorate voted to fund the plant's construction. After a $1.5 million turbine was purchased and four thousand feet of penstock were laid, the project was halted by a lawsuit. Among the complainants were several landowners with multimillion-dollar properties that abutted or drew water from Castle Creek. One of these, cloaked by four limited-liability companies that he controlled, was Bill Koch.

Bill was a pauper in comparison to his older brother Charles and his late twin brother, David, who together had donated more than one hundred and twenty-five million dollars to organizations that promoted global warming denial. Bill was worth only $1.8 billion. He had founded the energy conglomerate Oxbow Carbon, which operated a coal mine in a small town called Somerset, on the other side of the Elk Mountains, and a gas plant a few miles away. Some billionaires settled for private islands; Koch owned a private valley.

Just outside Somerset—and a mere fifteen-minute helicopter ride from his twenty-eight-bedroom Aspen palazzo—Koch had built Bear Ranch, which was about twice as large as the city of Aspen. Within the ranch stood a replica Old West town that "Wild Bill," as he was known to friends, built for his personal amusement. Most of the buildings had been imported, piecemeal, from Buckskin Joe, an MGM set used as a backdrop for *True Grit* and *Cat Ballou*. The town had seventy buildings, among them five saloons, a livery stable, a church, a bank, a theater, a hotel, a brothel, and a jail. A staff of twenty oversaw Koch's collection of more than one million items of memorabilia: Sitting Bull's rifle, General Custer's flag, Jesse James' gun, the only photo taken of Billy the Kid. Koch described the ranch as "a place where I can enjoy life and enjoy my family and friends without having to worry about my enemies." One of these enemies was an Oxbow Carbon executive named Kirby Martensen who claimed that after he was accused of fraud, he was kidnapped by Koch, brought to Bear Ranch, and interrogated. (His lawsuit against Koch was later dismissed.)

Another enemy of Koch's was the mayor of Aspen, Mick Ireland. "There are a lot of obstacles to renewables in this town," said Ireland, "that are not apparent to the naked eye."

Koch and the other opponents of the hydroelectric plant warned that it would cost taxpayers too much money and argued various technical legalities. But they found that their most successful strategy was to appeal to the anxieties of the city's environmentalists. They claimed that the plant, despite the assurances of the city's ecological consultants, would threaten the creek's survival. (Koch had used the same tactic in his five-million-dollar campaign to defeat an offshore wind farm in Nantucket Sound that would be visible from his summer compound; he said he worried the turbines would "destroy a beautiful environment.") Flyers showing images of dry streambeds and polluted water were sent to every Aspen mailbox. IT'S NOT GREEN TO KILL A STREAM, they read. ASPEN'S WATER-GATE. The ads were paid for by shell organizations with names like "the Aspen Citizens Committee" and "Friends of Castle and Maroon Creeks," registered in Denver and Colorado Springs.

The plant debate clawed open an intergenerational rift in Aspen's environmental movement. It pitted, as one hydro supporter put it, the "Rachel Carson type," who prioritized the protection of the local rivers and streams, against the "Auden Schendler type," who pointed out that every creek would dry up if radical measures weren't taken to curb atmospheric warming. "Nobody wants to be against Auden Schendler," said Chelsea Congdon Brundige, a local waterways activist. "He's the environmental playboy." But she warned that the stream would become ever more vulnerable as temperatures rose, which was a reason to protect it vigilantly. Besides, Brundige argued, Aspen's quest to ban coal-fired power was largely symbolic; it would have no appreciable impact on the global climate. The same, of course, could be said about the Aliso Canyon gas leak, the appliances running in Aspen's empty houses, or most any other single source of greenhouse gas emissions.

"It's an odd debate in many ways," said William Dolan, the city's utilities specialist. "Both sides could describe themselves as extremely environmentally conscious, and they'd be right. The vast majority of people on the other side of the issue had pure intentions. Whether their concerns were founded in reality is another question." But environmentalists had differing realities. This was because they had different pasts. The older generation had come of age defending isolated, pristine natural wonders against the predations of industry; the younger generation worried about the catastrophic disruption of the governing systems of life on Earth. The two generations also had different futures. The older activists worried about how the creek might look in twenty years. The younger activists worried about whether the creek would exist in fifty years.

The decisive blow came when a petition, funded by Bill Koch, collected enough signatures to require a new referendum. Voters expressed their disapproval of the hydro plant by a margin of one hundred and ten votes. The city suspended its plans; in the years that followed, it bought wind and hydroelectric power from plants as far away as Kimball, Nebraska. Like nearly every luxury item in Aspen, the city's renewable energy was imported from somewhere on the other side of the mountains.

The lesson Schendler drew from the failed hydro plant had ramifications that extended beyond Aspen, even beyond the American West. "If we want to solve climate change," he said, "we're going to fracture alliances. We're going to do some difficult things. And those things are going to hurt." This, at least, was how Schendler justified his most surprising gambit of all: a partnership with Bill Koch.

After ten years of entreaties, Schendler managed to obtain

a meeting with Oxbow Carbon to discuss the possibility of capturing methane vented by its coal mine, Elk Creek, one of the nation's largest underground mines. Methane gas trapped in coal seams is released during mining; Colorado law required mines to vent the gas to protect workers and prevent explosions. That methane rose into the atmosphere, where, over a twenty-year period, it trapped radiation eighty-four times more efficiently than carbon dioxide. But methane can also generate energy. Schendler proposed outfitting Oxbow's mine with a methane-capturing system that would generate enough energy to power all of Skico's properties. It would also eliminate three times Skico's annual greenhouse emissions.

"Wait," said an Oxbow representative, raising his palm. "We don't believe methane is a pollutant," he said. "We don't believe coal is, either. We think burning coal is good for society."

Schendler paused. What could he say to win men like this over?

"Why," he finally asked, "did you take this meeting?"

"I'm a resource guy," said the miner. "I hate to see resources wasted."

"It was the one, small piece of common ground," Schendler later said, "and I had to cling onto it for dear life."

The methane-capturing plant opened on November 9, 2012, three days after Aspen's voters passed the Koch-backed referendum to block construction of the hydroelectric plant. At the grand opening, Schendler posed for pictures with Oxbow executives. Koch declined to attend but he did provide a guarded statement for the press release. "This project," he said, "is useful and rational." It was not quite a ringing endorsement, though it did make Schendler's point as concisely as possible.

A few weeks later a tunnel deep in the mine burst into flames. The fire could not be extinguished. After a year, Elk Creek, responsible for a tenth of Colorado's coal production, shut down. Oxbow buried tens of millions of dollars of equipment underground. Koch laid off nearly three hundred employees, many of them third-generation miners.

Meanwhile the power plant thrived. Methane continued to leak from the sealed mine and was converted into energy—enough to power seventeen restaurants, two luxury hotels, and four ski resorts. "We're sorry for the job losses Oxbow is experiencing," Schendler told a reporter, "but we're glad that our project is still running."

The collapse of the coal mine had made Schendler's point too—a point about a new industry's cannibalization of an old one, about the violent end of a familiar way of life and the birth of another.

Schendler was named after the poet. His mother, who lived in New York City, often saw W. H. Auden in the subway, shuffling down the platform in bathrobe and slippers. Schendler's favorite poem was "Musée des Beaux Arts," which Auden wrote after viewing *Landscape with the Fall of Icarus*, attributed to Pieter Bruegel the Elder, at the Royal Museums of Fine Arts of Belgium. The painting is a view of a harbor, busy with ships, seen from a seaside cliff. In the foreground, on a cliff, a plowman tends to his crops; a shepherd stands with his flock; an angler casts his rod. Down in the lower right-hand corner, easily missed at first glance, are the pale white legs of Icarus, kicking as he crashes headfirst into the water.

"The point," said Schendler, "is that suffering happens on the sidelines, while life is being lived elsewhere. While we're in Aspen, a lot of bad things are going on in the world. We can be detached from all of that because it's so beautiful here."

It seemed a purposefully modest reading of the poem. For Schendler had less in common with the oblivious laborers than with Icarus. It was Schendler, after all, whose forsaken cry was ignored by the busy plowman and the passengers on the ship who, despite having seen "Something amazing, a boy falling out of the sky, / Had somewhere to get to and sailed calmly on." As Schendler cultivated his own dreadful martyrdom, setting a golden example even as Aspen's snow melted, would the world notice? Or would it sail calmly on?

Schendler had dedicated his life to the hope that his actions, however symbolic, would exert profound global influence. "There's a temptation to think we're the center of the universe, and we're not. We're a bubble. Lose that awareness and you're doomed. We're happy to say how we fucked up. And we've fucked up, don't get me wrong. We've *fucked up*. But we have the ability, in Aspen, to reach the world."

Schendler stood halfway up Fanny Hill, the beginner ski slope at the base of Snowmass Mountain. At the side of the trail, a small shack loudly hummed. Behind it a pool of turquoise water, a retention pond for snowmelt, was filled by a pipe that drained from West Brush Creek some eight hundred feet up the mountain. During the winter a machine turned this water into snow. Early in his tenure Schendler realized that the snowmaking system, one of the major drivers of Skico's energy consumption, could itself be used to generate electricity. For about two hundred thousand dollars, Skico built the powerhouse and installed within it a turbine. This micro-hydroelectric plant generated enough energy to power fifteen homes year-round. But the plant's long-term financial benefits meant little to Schendler. Nor did the energy savings, which, in the broader scheme of things, were infinitesimal.

"None of that matters," said Schendler. "But *this* matters." He pointed to a large informational sign attached to the side of the powerhouse. A diagram showed how the turbine, with the help of a Pelton wheel, a waterwheel invented by a gold miner, transformed water into energy.

Seven hundred thousand people, many of them obscenely rich, skied down Fanny Hill each winter. Some of these skiers, Schendler figured, would take a break at this plateau, or pause out of curiosity, or fall down, and read the sign. Some of those people, just maybe, would be encouraged to finance hydroelectric plants, or more ambitious projects, of their own.

But nobody was looking at the sign just then, because it was summer and there was no snow on the ground, only grass and rocks. In the stillness of the afternoon, it was difficult to imagine how the slope looked when it was a soft, brilliant white, plowed by thousands of skiers practicing snowplows and J-turns. It was much easier to picture a future in which the mountain was bare year-round—when nobody would have the chance to read the sign except, perhaps, the occasional hiker, trying to escape the heat of the valley.

Part III

AS GODS

7

PIGEON APOCALYPSE

The first time Ben Novak saw a passenger pigeon, he fell to his knees and remained there, speechless, for twenty minutes. He was sixteen years old. At thirteen, Novak had vowed to devote his life to resurrecting extinct animals. At fourteen, he saw a photograph of a passenger pigeon in an Audubon Society book and "fell in love." There was a single passenger pigeon in his home state of North Dakota but it was locked in the research collection of the historical society in Bismarck. He had no idea that the Science Museum of Minnesota, which he was visiting with a summer program for high school students, had pigeons in its collection, so he was stunned to encounter two stuffed specimens, a male and a female, mounted in lifelike poses. He felt overcome by awe, sadness, and euphoria at the sight of the birds' beauty: their bright auburn breasts, their slate-gray backs, and the dusting of iridescence around their napes that, depending on the light and angle, appeared purple, fuchsia, or green. Before his chaperones dragged him out of the room, Novak snapped a photograph with his disposable camera. The flash was too strong, however, and when the film was processed several weeks later, he was haunted to find that the photograph came back blank, just a flash of white light.

Since then, Novak has visited four hundred and eighty-three passenger pigeons. He has seen them at the Burke Museum in Seattle, the Carnegie Museum of Natural History in Pittsburgh, the American Museum of Natural History in New York, and Harvard's Ornithology Department, which possesses one hundred and forty-five specimens, including eight pigeon corpses preserved in jars of ethanol, thirty-one eggs, and a partially albino pigeon. There are 1532 passenger pigeon specimens left on Earth. They have all been dead since the afternoon of September 1, 1914, when a captive pigeon named Martha expired at the Cincinnati Zoo. She had outlasted George, the penultimate survivor and her only companion, by four years. As news spread of the species' imminent extinction, Martha became a minor tourist attraction. In her final years, whether depressed or just old, Martha, apart from an occasional tremor, barely moved. Zoo visitors threw fistfuls of sand at her to elicit a reaction. When she died, her body was taken to the Cincinnati Ice Company, frozen in a three-hundred-pound ice cube, and sent by freight train to Washington, D.C. At the Smithsonian Institution she was stuffed and mounted and visited, ninety-nine years later, by Ben Novak.

The fact that we can pinpoint the death of the last passenger pigeon is one of many peculiarities that distinguish the species. Thousands of species go extinct every year, but we tend to be unaware of their passing, because we're unaware of most species' existence. Since the publication of Carl Linnaeus' *Systema Naturae* in 1735, approximately 1.3 million species have been identified, not counting microbes. A 2011 estimate of 8.7 million living species by the biologist Boris Worm was greeted with skepticism. But the actual number might be an order of magnitude higher. As Stewart Brand, whose Long Now Foundation failed to compile a database of all living species, has written, "We're so ignorant, we don't know how ignorant we are."

The extinction of the passenger pigeon, together with the precipitous decline of the American bison, helped to inspire the creation of national wildlife preserves, early scientific studies of the continent's endangered species, and the passage of the first major conservation laws. Before Martha's death, despite increasing warnings from a small community of naturalists, animals tended to be viewed as an inexhaustible commodity. If extinctions had occurred, they were due to orderly processes of nature, not man. Near the end of the eighteenth century, most Americans didn't even believe in extinction. "Such is the œconomy of nature," wrote Thomas Jefferson, who was convinced that mammoths still existed on the continent, "that no instance can be produced of her having permitted any one race of her animals to become extinct; or her having formed any link in her great work so weak as to be broken."

The passenger pigeon's decline was impossible to ignore, however, because as recently as the 1880s it was the most populous vertebrate in North America. As many as four out of every ten birds were passenger pigeons. In *A Feathered*

River Across the Sky, Joel Greenberg estimates that the species' population "may have exceeded that of every other bird on earth." Migrating flocks would, in the words of a Dutch visitor to Manhattan in the 1620s, "shut out the sunshine" for hours, even days at a time. "The most fervid imagination cannot conceive their numbers," wrote the explorer Zebulon Pike in 1806, encountering a roost on islands near St. Louis. "Their noise in the wood was like the continued roaring of the wind, and the ground may be said to have been absolutely covered with their excrement." In 1813, in Kentucky, the ornithologist Alexander Wilson observed a flock that he estimated to contain 2,230,272,000 birds. In 1860, in Fort Mississauga, Ontario, the English naturalist W. Ross King watched from dawn to dusk as a mass of pigeons a mile wide, and at least three hundred miles long, flew overhead; stragglers followed for days. That flock, he calculated, contained 3,717,120,000 pigeons. By comparison, there are currently two hundred and sixty million pigeons in existence; New Yorkers may be surprised to learn that there are but one million pigeons in New York City. A single passenger pigeon nesting ground once occupied an area as large as eight hundred and fifty square miles, or thirty-seven Manhattans.

When a flock landed in a wooded area, you could not see the trees for the birds. Observers describe "small pyramids" or "apparent haystacks" that, when approached, revealed themselves to be elms or willows "completely loaded down with live birds." But you could smell them a mile away. The flocks left behind "thousands of wagon loads" of excrement that flattened the understory and trees "as if they had been girdled." And the birdsong . . . the pigeons did not coo so much as shriek. An approaching flock, in the words of a Wisconsin

hunter in 1871, made "a roar, compared with which all previous noises ever heard are but lullabies, and which caused
more than one of the expectant and excited party to drop
their guns, and seek shelter behind and beneath the nearest
trees. The sound was condensed terror."

But most hunters did not drop their guns, not when
a single shot fired into the sky could kill as many as one
hundred and thirty-two birds—a record achieved by a Canadian hunter in 1662. The species' abundance was a seduc-
tive enticement to mass slaughter. The pigeons were hunted
for their meat, which was sold by the ton (at the high end
of the market, Delmonico's served pigeon ballotines stuffed
with foie gras and truffles); for their oil and feathers; and for
sport. Shooting competitions claimed tens of thousands of
birds per contest; there is an account of a San Antonian killing four hundred and seventy-five pigeons with a stick, of the
capture and slaughter of seven hundred thousand birds at a
nesting in Michigan, of a pigeon dealer in Wisconsin shipping
two million birds to market in one year. Even so, the species'
decline—from approximately five billion to extinction within
a generation—baffled most Americans. *Science* published an
article claiming that the pigeons had all fled to the Arizona
desert. Others hypothesized that they had taken refuge in
the Chilean pine forests, on an island east of Puget Sound,
or in Australia. Another theory held that every passenger
pigeon had joined a single megaflock and disappeared into
the Bermuda Triangle.

Stewart Brand, who was born in Rockford, Illinois, in
1938, never forgot the mournful way his mother spoke about
passenger pigeons. During summers the Brands vacationed
near the top of Michigan's mitten, not far from Pigeon River,

one of the hundreds of American places named after the species. (Michigan alone has four Pigeon Rivers, four Pigeon Lakes, two Pigeon Creeks, Pigeon Cove, Pigeon Hill, and Pigeon Point.) Old-timers told stories about the pigeon that to Brand assumed a mythic quality. The beating of the birds' wings, they told him, sounded like Niagara Falls.

Some years ago Brand invited the zoologist Tim Flannery to speak at Long Now's Seminar About Long-Term Thinking, a monthly series held in San Francisco. They titled their conversation "Is Mass Extinction of Life on Earth Inevitable?" In the question-and-answer period, Brand, grasping for a silver lining, mentioned a novel approach to ecological conservation that was gaining wider public attention: the resurrection of extinct species, like the woolly mammoth, aided by novel genomic technologies developed by the Harvard molecular biologist George Church. "It gives people hope when rewilding occurs," said Brand, "when the wolves come back, when the buffalo come back." He paused. "I suppose we could get passenger pigeons back. I hadn't thought of that before."

Brand became obsessed with the idea. Reviving a vanished species was exactly the kind of ambitious, interdisciplinary, loopy project that he had championed as editor of the *Whole Earth Catalog*, which he founded in 1968 and edited until 1984, an encyclopedic compendium of tools and practices and diagrams and product placement that became a totem of his generation—a sacred text, found on coffee tables everywhere (or at least in metropolises and college towns), of the environmental movement that flowered in the 1970s. The catalog, Brand has said, "encouraged individual power"; its success, in any case, gave Brand more power than most individuals, allowing him intimate access to the world's most imaginative thinkers and patrons wealthy enough to finance those thinkers' most brazen plans. Since 1996, some of those ideas have materialized under the aegis of the Long Now Foundation, which Brand established to support projects designed to inspire "long-term responsibility." The term "Long Now," coined by Brian Eno (a founding board member), denotes the period spanning from ten thousand years in the past to ten thousand years in the future. Underlying Long Now's mission was the premise that one's empathy expands with one's temporal horizon. Human beings like to flatter themselves as being a long-term species, but we are, at best, middle term and act accordingly, treating distant generations to come with the anomic cruelty of psychopaths. As Eno put it,

> Our empathy doesn't extend far forward in time. We need now to start thinking of our great-grandchildren, and their great-grandchildren, as other fellow-humans who are going to live in a real world which we are incessantly, though only semi-consciously, building. But can we accept that our actions and decisions have distant consequences, and yet still dare do anything? . . . If we want to contribute to some sort of tenable future, we have to reach a frame of mind where it comes to seem unacceptable—gauche, uncivilized—to act in disregard of our descendants.

These sentences were written in 1995—or, in Long Now's preferred nomenclature, 01995—then the warmest year on record, when a chunk of ice larger than Puerto Rico broke

off the Antarctic Peninsula. Anyone capable of feeling gen-
uine empathy with one's unborn great-grandchildren could
only recoil in horror at our continuing consumption of fossil fuels, but long-term thinking also forced a revaluation of
conventional wisdom on nuclear waste storage, translation
dictionaries, artificial intelligence, digital archives, public land
use, space exploration, investment strategies, infrastructure,
philanthropy, and the appreciation of art.

The most literal proof of Long Now's concept was a three-
hundred-foot-tall clock designed to tick uninterruptedly for
the next ten thousand years, financed by a forty-two-million-
dollar investment from Jeff Bezos, and built inside an excavated mountain that Bezos owns in West Texas. There were
also more than two hundred acres of land in eastern Nevada
used by an ongoing climate monitoring research station,
and a disk of pure nickel inscribed with fifteen hundred lan-
guages and mounted on the *Rosetta* space probe, which in
2016 landed on Comet 67P/Churyumov-Gerasimenko, five
hundred million miles from Earth.

Three weeks after his conversation with Tim Flannery,
Brand sent an email to Church and Edward O. Wilson:

> Dear Ed and George . . .
> The death of the last passenger pigeon in 1914
> was an event that broke the public's heart and
> persuaded everyone that extinction is the core of
> humanity's relation with nature.
> George, could we bring the bird back through
> genetic techniques? I recall chatting with Ed
> in front of a stuffed passenger pigeon at the
> Comparative Zoology Museum [at Harvard, where

Wilson is a faculty emeritus], and I know of other stuffed birds at the Smithsonian and in Toronto, presumably replete with the requisite genes.
Surely it would be easier than reviving the woolly mammoth, which you have espoused.

The environmental and conservation movements have mired themselves in a tragic view of life. The return of the passenger pigeon could shake them out of it—and invite them to embrace prudent biotechnology as a Green tool instead of menace in this century . . . I would gladly set up a nonprofit to fund the passenger pigeon revival . . .

Wild scheme. Could be fun. Could improve things. It could, as they say, advance the story.

What do you think?

In less than three hours, Church responded with a detailed plan to return "a flock of millions to billions" of passenger pigeons to the planet.

On February 8, 2012, Church hosted a symposium at Harvard Medical School called "Bringing Back the Passenger Pigeon." It was attended by ten people, including Brand and Brand's wife, Ryan Phelan, an entrepreneur who founded an early consumer medical-genetics company; three ornithologists; and Beth Shapiro, an evolutionary molecular biologist who had already begun to sequence the passenger pigeon's genome in her lab at the University of California, Santa Cruz. Church gave a demonstration of his new genome-editing technology. "De-extinction went from concept to potential reality right before our eyes," said Phelan. "We realized that we could do it not only for the passenger pigeon, but for other species. There was so much interest that we needed to create an infrastructure around it. It was like, 'Oh, my God, look at what we've unleashed.'" Phelan became executive director of the new project, which they named Revive & Restore.

Several months later, the National Geographic Society held a debate of the scientific and ethical questions raised by the prospect of "de-extinction." Brand and Phelan invited three dozen of the world's leading genetic engineers and biologists, among them Stanley Temple, a founder of conservation biology; Oliver Ryder, director of the San Diego Zoo's Frozen Zoo, which stockpiles frozen cells of endangered species; Henri Kerkdijk-Otten, who was breeding European cattle in the hope of creating a new species that resembles the auroch, a form of wild ox, extinct since 1627, that is the ancestor of domestic cows; Mike McGrew, a biologist in Scotland who created a chicken that gives birth to ducks, and a duck that gives birth to chickens; and Sergey Zimov, who built an experimental preserve in Siberia called Pleistocene Park, which he hoped one day to populate with woolly mammoths. Steven Spielberg, then producing the fourth *Jurassic Park* film, sent a researcher to take notes.

To the argument that the pigeon project would provide "a beacon of hope for conservation," Brand added an ethical motivation: "Human beings have made a huge hole in nature. We have the ability now—and maybe the moral obligation—to repair some of the damage." This reasoning reflected the tacit belief in a communal human consciousness, extending from the nineteenth-century hunters who slaughtered millions of birds to conservationists born nearly a century after Martha's death, and with it a sense of guilt, unbounded by era or personal experience, that demanded of our species a profound act of expiation. Brand offered a vision of humanity as creation's moral authority—a fallen authority that had earned, through scientific progress, a chance for redemption.

The other conference attendees were more comfortable making narrower ecological claims in favor of de-extinction. Just as the loss of a species diminishes the richness of an ecosystem, with a farrago of unintended consequences, the addition of new animals might achieve the opposite. "It's not just about one species," said Phelan. "It's about that cascade effect that will happen to an ecosystem." The grazing habits of mammoths, to take one example, exposed the soil to cold air, which helped to protect the Arctic permafrost from melting—a benefit with global significance, as the permafrost contains two to three times as much carbon as the world's rain forests. It was important, Brand grasped immediately, to frame the project in terms of conservation. "We're bringing back the mammoth to restore the steppe in the Arctic," he said. "One or two mammoths is not a success. One hundred thousand mammoths is a success."

A less scientific, if more persuasive, argument was advanced by the ethicists Hank Greely and Jacob Sherkow. De-extinction should be pursued, they argued in *Science*, because it would be really cool. "This may be the biggest attraction and possibly the biggest benefit of de-extinction. It would surely be very cool to see a living woolly mammoth."

Ben Novak needed no convincing. When he heard that Revive & Restore had decided to resurrect the pigeon, he sent an email to Church, who forwarded it to Brand and Phelan. "Passenger pigeons have been my passion in life for a very long time," wrote Novak. "Any way I can be part of this work would be my honor."

Jakob Novak, born in Moravia, might have traveled through vast flocks of passenger pigeons when he migrated west to the Great Plains with his family in 1873. Jakob's son, Anton, moved to the Badlands of North Dakota in 1914, the year that Martha died. Anton was a farmer; his son, Anton Junior, opened an auto repair shop and raised geese, ducks, and pigeons. He kept in his living room a telescope that he trained on his bird feeder, not thirty feet away. He'd focus the lenses on the birds and show them to his grandson, Ben. "You could see the auricular feathers better than if you were examining the bird in your own hand with a magnifying glass," recalled Novak. "He loved every detail of them."

Novak grew up in a house built by his father three hundred feet above the Missouri River. The property stood halfway between Williston, the ninth-largest city in North Dakota, and Alexander, which has a population of two hundred and twenty-three. Three miles separated the Novaks from their closest neighbor. Two hours south was Elkhorn Ranch, where Theodore Roosevelt developed the theories about wildlife protection that led to the preservation of nearly a quarter billion acres of public land. As a boy Novak took solitary hikes though the Badlands, exploring a vast petrified forest that runs through the Sentinel Butte formation. Fifty million years ago, this part of western North Dakota was a lush forest, banded by rivers and swamps, its trees standing more than one hundred feet tall. It looked something like the Florida Everglades. Novak frequently came upon vertebrae, phalanges, and rib fragments of extinct crocodiles and champsosaurs. He spent a summer vacation digging up a bison skeleton.

The local schools emphasized conservation in their science classes. In the third grade, Novak was assigned school projects about endangered species, and a few years later he learned in an article in *National Geographic* that he was living in the middle of the sixth mass extinction in Earth's history and that changes made by human beings to the composition of the atmosphere could kill off, by 2050, a quarter of the planet's mammal species, a fifth of its reptiles, and a sixth of its birds. "I felt a certain amount of solidarity with these species," he said. "Maybe because I spent so much time alone."

When, at thirteen, Novak learned about cloning, he became convinced "that the future of conservation was filling the holes left behind by extinction." That year he entered his science fair with a project explaining the technical challenges that had to be overcome to resurrect the dodo bird (while preparing the project, he learned that the dodo was a giant pigeon). He won top prize in his junior high division of the North Dakota State Science Fair. "This victory," he later wrote, "would go on to decide the course of my life."

After graduating from Montana State University in Bozeman, Novak applied to study under Beth Shapiro, who had begun to sequence passenger pigeon DNA in 2001, a decade before Brand had his big idea. She rejected him. "I appreciated his devotion to the bird," she later said, "but I worried that his zeal might interfere with his ability to do serious science." Novak entered a graduate program at the McMaster Ancient DNA Centre in Hamilton, Ontario, where he worked on the sequencing of mastodon DNA, but remained obsessed with passenger pigeons. He decided that if he couldn't join Shapiro's lab, he would sequence the pigeon's genome himself. He petitioned every museum he could find that possessed stuffed specimens. He was denied more than thirty times before Chicago's Field Museum sent him a tiny slice of a pigeon's toe. A lab in Toronto conducted the sequencing for a little more than twenty-five hundred dollars, which Novak raised from family and friends. He had just begun to analyze the data when he learned about Revive & Restore.

After Phelan hired Novak, Beth Shapiro offered him a desk at the UCSC Paleogenomics Lab, where he could witness the sequencing work as it happened. When people asked him what he did for a living, he said that his job was to resurrect the passenger pigeon.

Ben Novak was tall, guarded, and decorous in professional conversation, until the conversation turned to passenger pigeons—which conversations with Novak usually did. He reserved laughter not for mirth but to express disbelief, as when he was asked whether de-extinction might turn out to be impossible. The punch line was that, by one definition, it had already happened. More than fifteen years ago a team of Spanish and French scientists led by Alberto Fernández-Arias (who would later advise Revive & Restore) resurrected a bucardo, a species of mountain goat also known as the Pyrenean ibex, that went extinct in 2000. The last surviving bucardo was a thirteen-year-old female named Celia. Before she died—her skull crushed by a falling tree—Fernández-Arias scraped skin off one of her ears and froze it in liquid nitrogen. (Celia was later stuffed and mounted at the reception center of the Ordesa y Monte Perdido National Park in Aragon, Spain.) Employing the same technology that created Dolly the sheep, the first cloned mammal, Fernández-Arias' team used Celia's DNA to create embryos that they implanted in the wombs of fifty-seven goats. One of the does successfully brought her egg to term on July 30, 2003. "To our knowledge," wrote the scientists, "this is the first animal born from an extinct subspecies." But it didn't live long. After struggling for breath for seven minutes, the kid choked to death.

This cloning method, called somatic cell nuclear transfer, can be used only on species for which we possess cellular material. For species like the passenger pigeon that had the misfortune of going extinct before the advent of cryopreservation, a more complicated process is required. The first step is to reconstruct the species' genome. This is difficult, because as soon as an organism dies, its DNA begins to decay. The DNA also mixes with the DNA of other organisms with which it comes into contact, like fungus, bacteria, and other animals. If you imagine a strand of DNA as a book, then the DNA of a long-dead animal is a shuffled pile of torn pages, some of the scraps as long as this paragraph, others a single sentence or just a few words—a polynucleotide Heraclitus. Worse, the scraps are not in the right order, and many of them belong to other books. And the story is an epic: The passenger pigeon's genome contains about 1.2 billion base pairs. If each base pair was a word, the saga of the passenger pigeon would run four million pages.

There is a shortcut. The genome of a closely related species will have a high proportion of identical DNA, so it can serve as a blueprint, or "scaffold." The passenger pigeon's closest genetic relative is the band-tailed pigeon, which Shapiro had already sequenced. By comparing the fragments of passenger pigeon DNA with the genomes of similar species, researchers could assemble an approximation of an actual passenger pigeon genome. How close an approximation, it will be impossible to know. As with any translation, there will be errors of grammar, garbled phrases, and omissions, but the book will be legible. It should at least tell a good story.

Novak offered a new name for this novel species: *Patagioenas neoectopistes*, or "new wandering pigeon of America." He also proposed a new definition of "extinction," arguing that any species for which we possess cryopreserved tissue, such as the bucardo, should be considered not extinct but "evolutionarily torpid." The goal of de-extinction, therefore, was the restoration of an ecosystem through the *adaptation* of living organisms, whether through breeding, artificial selection, or genome editing. Strictly speaking, de-extinct species would be not revived species but novel, man-made creatures.

In five years, Shapiro's team successfully sequenced the DNA of three dozen passenger pigeons. That was the easy part. Next the genome had to be recreated in living cells. The most efficient way to do this is in a petri dish, but pigeon cells are difficult to grow in the lab. The only bird species to have been successfully cultured is the domestic chicken. In 2019, Revive & Restore provided a grant to researchers at Texas A&M to determine how to modify the cells of other avian species. Once the process is perfected, Novak will begin editing band-tailed pigeon cells, swapping out chunks of DNA for synthesized passenger pigeon DNA. These mock–passenger pigeon embryos will be reimplanted into band-tailed cocks and hens, which will hatch birds that resemble, as closely as possible, passenger pigeons.

Here the de-extinction process will move from the lab to the coop. Applying the same methods used to breed endangered species in captivity, developmental and behavioral biologists will teach the pigeons to act like their extinct models. There may be a steep learning curve. Band-tailed pigeons are a western bird and migrate vast distances north and south; passenger pigeons lived in the eastern half of the continent and traveled randomly in search of food. Unlike any other pigeon species, passenger pigeons lived in vast colonies and were raised socially, by the group. Their parents abandoned them two weeks after birth. It was easy to do so, in a flock of millions, particularly when their habitats extended over thirty-seven Manhattans.

Novak will never abandon his pigeons; he will watch over them with exaggerated paternal pride. In large outdoor aviaries he will install acoustic speakers to play, in perpetuity, a din that simulates the sounds of million-bird flocks. Breeders will introduce other animals into the habitat, and transfer the birds from one aviary to another, to simulate their hopscotching migratory patterns. After ten years or so, Novak will set some of the pigeons free, monitoring them through GPS chips implanted subcutaneously. He will consider the project a success when the population in the wild fulfills the same ecological functions as the passenger pigeons did. The new wandering pigeons of America will be expected to revitalize forests through a campaign of mass destruction, their megaflocks collapsing branches and poisoning undergrowth with guano. Novak expected this would happen in his lifetime. The passenger pigeon project, he said in 2020, "is not technology bound. It is money bound. If I had a rich oligarch that gave us twenty-five million right now, I'd guarantee we'd have a passenger pigeon in five years." On his current trajectory, he figured the first passenger pigeons would be released into the wild in the 2050s.

His optimism had been boosted by the U.S. government's sudden enthusiasm for using genetic tools to advance conservation goals. One of the few species to suffer a collapse commensurate with that of the passenger pigeons was the American chestnut, which until the early twentieth century was the dominant tree of the nation's eastern hardwood forests. A blight introduced by the importation of Asian chestnut trees killed billions of American chestnuts within several decades; today only a few dozen exist in the wild. Novak expected that a novel transgenic American chestnut, engineered to resist the deadly fungus, would soon be planted in national parks. The Department of Agriculture had already approved virus-resistant plum and papaya trees. The U.S. Fish and Wildlife Service had taken more active involvement in one of Revive & Restore's own projects, the revival of the black-footed ferret, one of the continent's most endangered species. In 2013, Fish and Wildlife invited Revive & Restore to explore the use of genomic technologies to engineer resistance to sylvatic plague, which threatened the ferret's survival. The regulatory precedents were being established. Novak just needed the birds.

While they waited for the researchers at Texas A&M to figure out how to culture pigeon embryos, Revive & Restore monitored the world's other species-resurrection efforts. The Lazarus Project was trying to revive an Australian frog, extinct for thirty years, that gave birth through its mouth. Beth

Shapiro had taken tissue samples from Monarch, the California grizzly bear, dead for more than a century, who served as the model for the state flag. An Australian researcher who had sequenced the Tasmanian tiger's genome hoped to gestate a genetically modified embryo in the pouch of a kangaroo. There were plans to bring back the Carolina parakeet, Steller's sea cow, and the great auk, which hasn't been seen since June 3, 1844, when the last two known members of its species were strangled on the island of Eldey by Icelandic fishermen, one of whom, on the way out, smashed an auk egg beneath his boot.

De-extinction was, even by the standards of an organization that thought in deca-millennia, a slow process. Every year the genetic engineering technology accelerated exponentially, but still no extinct species were close to reanimation. Ryan Phelan, recognizing that the same technology that could bring back extinct species could save endangered ones, broadened Revive & Restore's mission. Genetic techniques could give sea stars immunity to wasting disease, develop microbes to break down forever chemicals like PFOA, breed saltwater-resistant cypress trees to defend disappearing coastal wetlands. The scientific term for this type of genetic intervention is "facilitated adaptation." A better name for Revive & Restore would be Revive & Restore & Improve.

The optimistic, soft-focus fantasy of de-extinction, however thrilling to Ben Novak, disturbed many conservation biologists, who considered it a threat to their entire discipline and even to the environmental movement. At Revive & Restore conferences and in the popular and academic press, they articulated their litany of criticisms at an increasingly high pitch. In response, supporters of de-extinction advanced ever more aggressive counterarguments. "We have answers," said Novak, "for every question."

The most pressing question was practical: money. Environmental philanthropy was a competitive market. The conservation movement's sales pitch was heavily dependent on what biologists referred to as "charismatic" species, the animals that populated children's books—koalas, butterflies, rhinos, gorillas, polar bears—to draw donors. De-extinction was a flashy new rival for conservation's oldest philanthropic lure. What was more charismatic than a woolly mammoth? As the conservationist David Ehrenfeld said, at a Revive & Restore conference, "If it works, de-extinction will only target a very few species and is extremely expensive. Will it divert dollars from tried-and-true conservation measures that already work, which are already short of funds?"

Brand and Phelan dismissed this worry out of hand by lamenting their own difficulties raising money. Besides, they argued, any success they enjoyed would only drive more money to conservation. "We get asked these questions," said Novak, "but no one is asking the people who work on rhinos why they aren't working on the Arctic pollinators that are being devastated by climate change. The panda program rarely gets criticized, even though that project is completely pointless in the grand scheme of biodiversity on this planet, because the panda is a cute animal. Ultimately the great thing about conserving pandas is that it gets people who have never heard about conservation to think about it. The same is true of de-extinction."

De-extinction posed a deeper rhetorical threat to conservationists. The specter of extinction has been the conservation movement's most powerful justification for its existence. What if extinction began to be seen as a temporary inconvenience? As the ecologist Daniel Simberloff, a frequent collaborator with his mentor Edward O. Wilson, said, "Technofixes for environmental problems are band-aids for massive hemorrhages. To the extent that the public, who will never be terribly well informed on the larger issue, thinks that we can just go and resurrect a species, it is extremely dangerous . . . De-extinction suggests that we can technofix our way out of environmental issues generally and that's very, very bad."

Novak rejected this logic: "This is about an expansion of the field, not a reduction." If the success of de-extinction, or even its failure, increases public awareness of the threats of mass extinction, Novak argued, it will have been a triumph. In its first decade Revive & Restore had achieved only middling scientific success, but as a publicity campaign it had been nothing short of a juggernaut, a steady, reliable source of magazine covers, editorials, and popular science books.

Might a new bird invite a new bird flu? "Pathogens in the environment are constantly evolving, and animals are developing new immune systems," said Doug Armstrong, a conservation biologist in New Zealand who studied the reintroduction of species into the wild. "If you re-create a species genetically and release it, and that genotype is based on a bird from a hundred-year-old environment, you probably will increase risk."

David Haussler, a bioinformatician at UCSC who advised Revive & Restore, didn't buy it. "There's always this fear that somehow if *we* do it, we're going to accidentally make something horrible, because only nature can really do it right. But nature is totally random. Nature makes monsters. Nature makes threats. Many of the things that are most threatening to us are a product of nature. Revive & Restore is not going to tip the balance." Human beings have been in the business of making monsters for thousands of years. Cows were once aurochs; chickens were once red jungle fowl; pigs were wild boars. Nobody feels threatened by barnyard animals or, for that matter, municipal parks, yellow bananas, or cornfields. It is the wild species, those that have managed to evade our domination—the grizzly, the great white, the civet—that retain the power to terrify.

Was it not cruel to experiment on animals, even killing them, in a long-shot bet on species resurrection? Consider the fifty-six female mountain goats that were unable to bring to term the bucardo embryos implanted in their wombs. Or the cloned bucardo that lived only a few minutes, gasping for breath, before dying of a lung deformity. "Is it fair to do this to these animals?" asked Shapiro. "Is 'because we feel guilty' a good enough reason?"

Stewart Brand offered a utilitarian counterargument. "We're going to go through some suffering," he acknowledged, "because you try a lot of times, and you get ones that don't take. On the other hand, if you can bring bucardos back, then how many would get to live that would not have gotten to live?"

The most tenacious questions posed by conservationists addressed the logic of bringing back an animal whose native habitat has disappeared. Why go through the trouble just to have the animal go extinct all over again? And why a species known for its explosive diarrhea? As Beth Shapiro said, "Do you think that wealthy people on the East Coast are going to want billions of passenger pigeons flying over their freshly manicured lawns and just-waxed SUVs?"

Novak originally responded by arguing that the passenger pigeon should be especially adaptable because it was an opportunistic eater with no specific native habitat. Over time, however, his argument deepened, becoming the subject of his master's thesis, for which Shapiro served as adviser. The passenger pigeon, he concluded, did not merely benefit from the sprawling domain of the continent's unpeopled forests; it was the forests' architect. The pigeon was "*the* ecosystem engineer of eastern North American forests for tens of thousands of years, shaping the patchwork habitat dynamics that eastern ecosystems rely on." The destruction wrought by nomadic armies of pigeons—branches cracking beneath their weight, acres of ground cover carpet bombed by bird shit—stimulated biodiversity, much in the way that the "simplified" Lower Ninth Ward invited colonizing plants and animals. Novak likened the pigeons' devastation to "a storm and a wildfire all in one," creating ecological hot spots that simmered long after the flock had moved on.

Novak designed a four-year experiment to demonstrate this. He planned to stage it on two hectares of forest, separated by some distance but as similar as possible in ecological profile. Biologists would spend the first year establishing baselines: counting species, drawing soil samples, reading tree rings. The following spring, Novak and about fifty assistants would descend upon one of the plots, doing their best impersonation of a six-week pigeon roost. Novak and his band of droogs, brandishing saws and chains and cudgels—*britvas* and *oozies* and *shlagas*—would knock over trees, stomp on foliage, hurl grappling hooks over branches and pull. They would confiscate every acorn and berry in the hundred acres. They would dump three eighteen-wheeler truckloads of pigeon guano on the forest floor, install pigeon decoys on the treetops to alert predators, and strew the ground with pigeon carcasses. "You can preach all you want about the ecological benefits of passenger pigeons," said Novak, "but many people won't understand the significance until you show them the effect of a bunch of bird poop out in the environment." Stewart Brand named Novak's experiment "Pigeon Apocalypse."

The corollary to Novak's conclusion about the ecological benefit of megaflocks of passenger pigeons could be seen from miles away: to revive and restore our forests, the pigeons must return. The passenger pigeon, argued Novak, was "the most important species for the future of conserving eastern America's woodland biodiversity."

With this synthesis, Novak's mission in life came full circle. The extinct bird that had won his fourteen-year-old heart just happened to be the species best suited to save the forests of North America.

Much of the romance and skepticism roused by de-extinction was inspired by semantic confusion. The term "de-extinction," as the geneticists were the first to concede, was a figure of speech—a metonymy, to be precise. The International Union for Conservation of Nature, the global authority on endangered species, defines "de-extinction" as the creation of "proxies" of extinct species that serve the same ecological functions

but are not "faithful replicas." Passenger pigeons will not rise
from the grave. We won't know how closely the new pigeon
will resemble the extinct pigeon until it is born, and even
then we'll be able to compare only physical characteristics.
Our understanding of the passenger pigeon's behavior, after
all, derives entirely from historical accounts. While many of
these, including John James Audubon's chapter on the pigeon in *Ornithological Biography*—"the tenderness and af-
fection displayed by these birds towards their mates are in
the highest degree striking"—are vivid and touching, few
are scientific in nature. "There are a million things that you
cannot predict about an organism just from having its genome sequence," said Ed Green, a biomolecular engineer who
worked on genome-sequencing technology in the UCSC Paleogenomics Lab. "It's one guess," said Shapiro. "It's not even
a very good guess."

She was no more sanguine about the woolly mammoth
project. "You're never going to get a genetic clone of a mammoth. What's going to happen, I imagine, is that someone,
maybe George Church, is going to insert some genes into
the Asian elephant genome that make it slightly hairier. That
would be just a tiny portion of the genome manipulated, but
a few years later you have a thing born that is an elephant,
only hairier, and the press will write, 'George Church has
cloned a mammoth!'"

Church, though he planned to do more than alter the
hairiness gene, conceded the point. "I would like to have an
elephant that likes the cold weather," he said. "Whether you
call it a mammoth or not, I don't care."

Should anybody else? There is no authoritative definition
of "species." The most widely accepted definition describes

a group of organisms that can procreate with one another and produce fertile offspring. But there are many exceptions, including the one provided by "ring" species like the song sparrow, whose population ranges from southwestern Alaska to the eastern tip of Newfoundland. A member of a ring species can mate with peers in adjacent geographic areas—its "ring"—but not across the entire range. A Mexican song sparrow might not be able to mate with an Aleutian song sparrow, but they sing the same tune.

It was the same principle that informed artistic restoration: If you visit *The Last Supper* in the refectory of the Convent of Santa Maria delle Grazie in Milan, you will not see a single speck of paint from the brush of Leonardo da Vinci. You will see a mural with the same design as the original, and you may feel the same sense of awe as the refectory's parishioners experienced in 1498, but the original paint oxidized centuries ago. Should you visit Kyoto, your guidebook may describe Kinkaku, the Golden Pavilion, as a fourteenth-century shrine, even though the building has been destroyed and rebuilt many times. To those who worship there, the current Golden Pavilion is not a replica; it *is* the Golden Pavilion, and it has always been. Philosophers have named this phenomenon Theseus' Paradox, a reference to the ship that Theseus sailed back to Athens from Crete after he had slain the Minotaur. The ship was preserved by the Athenians who, as Plutarch writes, "took away the old planks as they decayed, putting in new and stronger timber in their place." Theseus' ship became a test for the philosophers, with "one side holding that the ship remained the same and the other contending that it was not the same."

What did it matter if Passenger Pigeon 2.0 was no more than a persuasive impostor? If the new, synthetically created bird enriched the ecology of the forests it populated—or even helped to create new ones—few conservationists or philosophers would object. The genetically adjusted birds would be no more novel to the deciduous forest ecosystem than other human interventions, most of them harmful: invasive species, disease, deforestation, and a rapidly warming climate. When human beings first arrived on the continent, they encountered camels, eight-foot beavers, and ground sloths as heavy as pigs. "People grow up with this idea that the nature they see is 'natural,'" said Novak, "but there's been no real 'natural' element to the earth the entire time human beings have been around."

Biologists have created new forms of bacteria in the lab, modified the genetic code of countless living species, designed transgenic creatures, and cloned dogs, cats, and water buffalo. But the engineering of self-propagating novel vertebrates—of breathing, flying, defecating pigeons— would represent a new milestone for synthetic biology. This was the fact that would overwhelm all arguments against de-extinction. Thanks, perhaps, to *Jurassic Park*, popular sentiment was already behind it. ("That movie has done a lot for de-extinction," said Stewart Brand in all earnestness.) In a recent poll by the Pew Research Center, half of the respondents believed that an extinct animal would be brought back by 2050. Among Americans, belief in the prospect of de-extinction trails belief in evolution by only ten percentage points. "Our assumption from the beginning has been that this is coming anyway," said Brand, his eye glimmering, "so what's the most benign form it can take?"

What is coming will extend far beyond the resurrection of extinct species. For millennia, we have customized our environment, our vegetables, and our animals, through fertilization, breeding, and pollination. Synthetic biology offers more sophisticated tools. Danny Hillis, a Long Now board member and prolific inventor who pioneered the processing mechanism that is the basis for most supercomputers, called de-extinction merely "the most conservative, earliest application of this technology." He was reminded of Marshall McLuhan's observation that the content of a new medium is the old medium: that each new technology, when introduced, recreates the familiar technology it will supersede. The earliest television shows were filmed radio shows. The first computers imitated typewriters. Synthetic biology, in the same way, may gain widespread public acceptance through the resurrection of charismatic animals. "Using the tool to re-create old things," said Hillis, "is a much more comfortable way to get engaged with the power of the tool." The first wave of cultured meat will mimic chicken nuggets; the first wave of de-extinction will target species that evoke a yearning for some purer—and imaginary—lost world. Only then will they be followed by a profusion of hybrid species, designer species, hyperintelligent species.

It will be necessary to get comfortable with the tool of synthetic biology because, as McLuhan put it, "once any new technology penetrates a society, it saturates every institution of that society." The opening line of the first *Whole Earth Catalog* was "We are as gods and might as well get good at it." Brand has since revised this motto to "We are as gods and HAVE to get good at it." (The all caps are his.) De-extinction was good practice.

Brand predicted that Revive & Restore's current work would soon seem quaint. "People will say, woolly mammoths with just two tusks, that's all you're doing? Not four?" For millennia humanity has strived, with ever greater aggression and brutality, "to batter nature into submission. This is a whole different approach: more humble and more adroit. The skill we're learning is how to *nuance* nature." Brand and Phelan often invoked the word "mythic" when describing the work of Revive & Restore. "It's a kind of storytelling that we're engaged in," said Brand. "Once you're in the thick of the story, you start to take the long term seriously." Seen from this angle, Revive & Restore was less a science project than a work of art.

Ben Novak, for his part, insisted that he was driven purely by ecological concerns. The scientists who worked beside him in the Paleogenomics Lab—witness to his daily passenger pigeon rhapsodies—suspected a second motivation. "Look, I'm a biologist; I've seen people passionate about animals before," said André Soares, a soft-spoken Brazilian biologist who spent much of his time analyzing passenger pigeon DNA. "I meet people who love birds, or small obscure rats. But I've never seen anyone this passionate." He laughed. "It's not like he ever saw the pigeon flying around. And it's not like a dinosaur, a massive beast that walked around millions of years ago. No, it's just a pigeon. I don't know why he loves them so much. I'm pretty sure he doesn't know either. What does he say?"

Novak said that the pigeon project was "all under the framework of conservation."

"I'm not sure he's being honest with himself," said Soares, who had never bothered to visit a stuffed pigeon. "I think the birds are his thing."

Ed Green, the biomolecular engineer down the hall, was more succinct. "The passenger pigeon," he said, "makes Ben want to write poetry."

8

BAYOU BONJOUR

I. Oil and Gas Is the Fabric of Your Town

The red seaweed was out of a nightmare, crackling underfoot like the bones of small birds. It was everywhere, obscuring the flat whiteness of the beach on Grand Isle, Louisiana's southernmost barrier island. Nobody in Grand Isle had ever seen anything like it. Each morning concerned residents swept the seaweed into garbage bags, but when the tide went out, the seaweed returned, a maroon blanket that lapped onto the beach like a giant rotten tongue. Why was it red? Why was there so much of it? No one knew.

Mike Stagg walked across the seaweed to fill a pair of two-liter bottles with ocean water for a political stunt he had planned. The idea was to carry the water to the governor's mansion in Baton Rouge, a distance of one hundred and fifty miles, by foot. Once there he would ceremoniously dump the water onto Bobby Jindal's front porch. Stagg wanted to deliver a message: The ocean is coming for us. If Jindal refused to save the dying coast, they would bring the coast to him.

Stagg thought that it would be a fun, whimsical way to get attention—an environmental freedom march. He expected that as he walked across southern Louisiana, giving radio and television interviews along the way, responsible citizens would join his pied piper's song.

Seven people, including one driving a support truck, began the march with Stagg in Grand Isle, at the origination of Louisiana Highway 1. It was June of what was then the hottest summer in recorded history. The road had no shoulder and the situation immediately became perilous. Cars seemed to accelerate when they saw the activists, forcing them to straddle the thin, uneven fringe of grass that divided the road. The water jugs were heavy. Stagg realized, too late, that he should have brought more socks. Before they'd made ten miles, everybody had given up except for Stagg, who was sixty-two, and his friend Kirk Green, a forty-year-old middle school history teacher in East Baton Rouge. They decided to divide the walking: one would drive the support truck ahead and wait, while the other carried the water bottles. They switched places every mile, leapfrogging each other. By noon each day it felt hotter than a hundred degrees.

On the second day, after more than thirty miles, the marchers met in Golden Meadow a seventy-nine-year-old Houma Indian who had seen the land disappear beneath his feet. In his youth he had hunted deer and trapped rabbit. As salt water entered the marsh and the land subsided, the walking paths swallowed by the tide, he became an oysterman. Now the water was too salty for oysters.

By the third day Stagg and Green took to having five-hour lunch breaks, sheltering inside gas stations to avoid the worst of the heat. They began to doubt whether anyone would join

the march. Their shoes were badly scuffed, their necks perpetually lathered with sweat. At a truck stop in Labadieville,
ninety miles on, they hunched over a Formica table chugging
bottled water.

"If it wasn't for the water of the state, its lakes and rivers,
we'd be speaking Spanish right now," said Green, the history
teacher. He traced his own lineage to the arrival of an ancestor named François Gautreau in 1632. "The only reason
the French wanted to found a colony here was because of
the Mississippi and the delta, with its rich resources. It's the
reason Jefferson bought it. If it wasn't for the water, there's
a possibility that we'd not even be part of America. Water
has been our livelihood going back thousands of years, to the
Houma Indians. We're losing our heritage and our culture.
You can't get that back. Oil is a finite resource. When the oil
is gone, the oil companies will be gone. But the Mississippi
replenishes."

"There will be an accounting," said Mike Stagg.

On the eighth day they reached the governor's mansion.
They were joined at last by a dozen activists waving homemade signs: HEALTHY WATER OVER PROFIT and BOBBY YOU
HAVE SUNK US. Two women strummed acoustic guitars and
harmonized protest hymns. The governor's security guards
refused to let the marchers get closer than the fence that surrounded the property. Stagg poured the plastic bottles of seawater on a ragged strip of grass between the fence and the
sidewalk.

"There will be an accounting," said Mike Stagg.

In football-mad Louisiana, the common way to visualize the
state's existential crisis is through the metaphor of football
fields. The state loses a football field's worth of land every
hundred minutes. That is an improvement: between 1932
and 2010 the rate was a football field every hour. The metaphor has had the inoculating effect of narrowing the devastation to a scale comprehensible to the human imagination.
As you expand the time horizon, however, the horror grows
surreal. Every two days, the state loses nearly the accumulated acreage of every stadium in the NFL. Each week, the
state loses the equivalent of every Division I football field in
the country. Were this rate of land loss applied to New York,
Central Park would disappear in a month and a half. Manhattan would vanish in a little more than two years. New
Yorkers would notice this kind of land loss. The world would
notice. But the hemorrhaging of Louisiana's coastal wetlands
went largely unremarked beyond state borders. This was peculiar, because the remaining three million acres of dying
marsh—approximately the landmass of Connecticut—are
the coast's first line of defense against the ouroboric perils
of hurricanes and sea level rise. The marsh defends nearly
one-fifth of the nation's crude-oil production; nearly ten
percent of its natural gas reserves; a terminal connected to
more than half of the nation's oil-refining capacity; the city
of New Orleans and its port; the homes of more than one and
a half million people; and the integrity of the lower Mississippi River, which conveys nearly forty percent of the nation's
agricultural exports. The attenuation of Louisiana, like any
environmental disaster carried beyond a certain point, is a
national-security threat.

Already the damage is plainly visible to airplane passengers arriving in New Orleans. The lowest part of the Mississippi River delta looks like a maple leaf devoured to its veins by cankerworms. Since 2011, the National Oceanic and Atmospheric Administration has delisted more than forty place-names from maps of the lower Mississippi. English Bay, Cyprien Bay, Skipjack Bay, and Bay Crapaud have merged like soap bubbles in a bathtub.

The land loss has swiftly reversed the process by which the state was built. As the Mississippi shifted its course over millennia, spraying like a loose garden hose, it deposited mud and silt in a wide arc. This sediment settled into marsh and over time thickened into land. But what took seven thousand years to create has nearly been undone in the last ninety. Dams built on the tributaries of the Mississippi, as far north as Montana, have reduced the river's sediment load by half. Levees penned the river in place, preventing floods from dispersing sediment. The dredging of two major shipping routes, the Mississippi River Gulf Outlet and the Gulf Intracoastal Waterway, invited salt water into the wetlands' atrophied heart. Beyond those scapular incisions, the oil and gas industry had hastened the marsh's death by a thousand cuts—or ten times that. The wetlands have been scored by ten thousand miles of canals, dug for tanker ships and pipelines. The canals drew salt water from the gulf, eroding the fragile root systems that held the wetlands together like woven thread. The drilling of fifty thousand oil and gas wells created air pockets that compounded the deterioration. By its own estimate, the industry has caused thirty-six percent of all wetlands loss in southeastern Louisiana. The Interior Department has calculated the industry's liability as high as fifty-nine percent.

A better analogy than disappearing football fields was proposed by the historian John M. Barry, who had lived in the French Quarter for about fifty years. Barry likened the marsh to a block of ice. The reduction of sediment in the Mississippi and the construction of levees "created a situation akin to taking the block of ice out of the freezer, so it begins to melt." Dredging canals and laying pipelines was "akin to stabbing that block of ice with an ice pick."

The oil and gas industry committed this damage with the tacit blessing of the federal and local governments and without substantial opposition from environmental groups. Oil and gas is, after all, Louisiana's dominant business, responsible for around two billion dollars in annual tax revenue. Industry executives had good reason to be surprised when, all of a sudden, they were asked to pay damages. The request came in the form of the most ambitious environmental lawsuit in the history of the United States. It was served by an unlikely antagonist, a former college-football coach, competitive weight lifter, and author of dense, intellectually searching five-hundred-page books of American history: John M. Barry.

When Hurricane Katrina made landfall in Louisiana on August 29, 2005, Barry was a year and a half into writing his sixth book, *Roger Williams and the Creation of the American Soul*, about the Puritan theologian's efforts to define the limits of political power. Barry was not a fast writer; his books took him, on average, eight years to complete. "I tend," he said, "to obsess." Earlier in his career, he spent nearly a decade as a political journalist, writing about Congress, an experience he drew upon for his first book, *The Ambition and the Power*, an *Advise and Consent* for the 1980s. Following its publication he quit journalism and cocooned himself in research, reading, and writing. He sought complex episodes of American history that in his rendering become Jacobean dramas about tectonic struggles for power. *The Ambition and the Power* would make an appropriate subtitle for any of his books—particularly *Rising Tide*, his history of the 1927 Mississippi River flood, the most destructive in American history.

In the days after Hurricane Katrina, Barry became one of the city's most visible ambassadors. "I felt I had an obligation," he said, "to convince people that the city was worth rebuilding." From the beginning Barry was careful to emphasize the oil and gas industry's responsibility for the devastation—a point rarely made in the national press. On the first Sunday after the storm, he published an editorial in *The New York Times* explaining that the canals that julienned the marsh below New Orleans had endangered the state's survival. On *Meet the Press*, he warned that there was "no point in building the levee system if you're not going to rebuild the coastline." When Dennis Hastert, the Speaker of the House, questioned the logic of coming to New Orleans' aid, Barry responded with an essay for *Time*, co-authored by Newt Gingrich, in which he pointed out that the United States spent more money every year in Iraq than it would cost to restore the entire Louisiana coast.

Meanwhile, like many other Americans, Barry was trying to figure out exactly how New Orleans had flooded so catastrophically. The numbers—the wind shear on Lake

Pontchartrain, the storm surge, the inches of rainfall—didn't add up. After making calls to his old sources, he concluded that the levees hadn't been overtopped, as officials from the U.S. Army Corps of Engineers assumed, but had collapsed because of design flaws. The needless deaths reminded him of the hundreds of innocents who died during the 1927 floods because of cynical decisions made by hidebound politicians. It confirmed to him that Katrina was not the story of a meteorological event; it was a story of politics, science, and power. It was, in other words, a familiar story.

Barry next made a decision that few writers in his position, presented with an obvious, urgent, and lucrative book idea, would make. He chose not to write a book. The subject was too personal, the outcome too uncertain. He spoke about his sense of obligation to the city, but there was more to it than that. The story of how a city—his city—had been ruined by the very forces that had shaped its identity had to be communicated to the rest of the country, Congress, and the president. Barry began to resemble the civil engineer Andrew Humphreys, a conflicted hero of *Rising Tide*, who arrived in New Orleans in 1850 to study flood control: "The work obsessed him, unbalanced him, pushed him to the margin. He stopped writing . . . because it distracted him . . . He himself talked to reporters. He basked in their attention, basked in their portrayal of him as a major figure." Barry put aside *Roger Williams* and, on the cusp of his sixtieth birthday, entered the arena. He joined state committees charged with rebuilding the coast. He toured flood-management systems in the Netherlands. In Washington, where he lived part of the year, he arranged meetings with lawmakers. Among these was a freshman congressman from the First District, which includes much of southeastern Louisiana: Bobby Jindal.

"We're getting nothing out of the White House," said Barry, in Jindal's office. "We're getting nothing out of the mayor. We're getting very little out of the governor." He hoped that Jindal, who had to abandon his home in the New Orleans suburbs during the storm, might be the hero that New Orleans needed. He begged the congressman to demand action from the White House. At the end of two hours, Jindal told Barry that taking a leadership position on Hurricane Katrina "didn't fit his timing for running for governor."

Barry left in disgust. "I knew everyone who ran for president in two cycles," said Barry, "and I would rate Jindal first in putting his personal ambition ahead of everything else."

Barry had a special talent for disgust. His face at rest conveyed a pained forbearance, as if bracing himself to correct some misapprehension or guileless hyperbole. Hardened by decades of studying the behavior of powerful people motivated by low ambitions and short-term thinking, he was quick to skepticism and impatient with ignorance. His voice was gravelly and low, though indignation amplified it fivefold. When he got hot, talking about his enemies, he impersonated a strafing machine gun. Barry was five feet nine and powerfully built; he participated in national weight-lifting competitions into his fifties. He played football at Brown (some twenty years before Jindal matriculated) and, after dropping out of a PhD history program, joined Tulane University's football team as a receivers' coach. In his office in downtown New Orleans, the most prominent wall decoration was a laminated *Times-Picayune* from 1973 celebrating the first Tulane victory over LSU in twenty-five years. His first paid writing assignment was an article for *Scholastic Coach* titled "Flexible Blocking Patterns," in which he explained the finer points of calling audibles at the line of scrimmage. When discussing his public battles, he summoned football metaphors. "Writing is pretty isolated," he said. "I enjoy the action. I like to fight."

In the post-Katrina miasma of incompetence and terror, Barry's combativeness and intellectual gravity elevated him rapidly in the public trust. When the Louisiana legislature created a regional levee board, Barry applied for a seat. (The levee board was technically the Southeast Louisiana Flood Protection Authority-East, or SLFPA-E, which, consistent with the New Orleanian flair for aberrant pronunciation, was referred to as "slip-fah.") The board oversaw flood-management projects in its jurisdiction, which included most of New Orleans. Because the city's continued existence depended on its ability to manage floods, this was no small responsibility. Barry was appointed in 2007 and served as the board's spokesman.

It seemed right that a historian should hold this position. Barry, like many historians, was fond of quoting George Santayana: "Those who cannot remember the past are condemned to repeat it." But he never quoted Santayana without quoting Hegel: "The only thing we learn from history is that we learn nothing from history." The battle for Louisiana's survival would be fought along the Santayana-Hegel axis.

Barry's conclusion that the marsh, not the levees, posed the more urgent existential threat had been reached about a year

earlier by a disgruntled former oil geologist named John Lopez. A Kris Kristofferson type with a charmingly surly demeanor, Lopez was the kind of indelicate truth-teller whose
prognostications earned grudging respect even from those
who believed his big ideas would doom them to penury and
ruin. He was born in New Orleans, one of five children, his
father a dealer of auto parts and, when they didn't sell, candy.
His mother was a human relations manager for the U.S. Army
Corps of Engineers. On weekends Lopez took family fishing
trips for speckled trout downriver in Plaquemines Parish
and for bull croakers at the edge of Lake Pontchartrain, on
New Orleans' northern shore. It was during his childhood,
in the 1960s, that the lake was judged too filthy for swimming. For decades it had served as a repository for untreated
sewage, agricultural discharge, and industrial waste, giving
its choppy waters the appearance, as one conservationist put
it, of "chocolate milk." The reputation survived the lake's removal from the EPA's Impaired Waterways list in 2006; the
local weather page still gives daily fecal coliform readings.

Lopez studied geology at LSU and went to work for
Amoco, mapping the seafloor of the Gulf of Mexico. While
prospecting for oil, he developed a heightened appreciation
of his industry's role in the despoliation of southern Louisiana. He began to call himself—in private—a "closet environmentalist." After moving to a property on the north shore,
he helped found the Pontchartrain Conservancy, dedicated
to improving the lake's water quality. The lake was not only
polluted; it also flooded regularly. Lopez realized that the
problems of wetland erosion that he encountered at Amoco
contributed directly to the increased flooding of his yard. Pon-
chartrain, after all, was not technically a lake but a gigantic

brackish estuary connected through a series of intermediary passes to the Gulf of Mexico. The lake could be only as healthy as the coast. "That," said Lopez years later, "was the revelation. I began to see the lake in a much broader context. In every historical period, we've always thought we're just *managing* our environment. But when you step back, you see that we have always been hell-bent on exploiting resources as quickly as we could."

After twenty years, Lopez quit Amoco and returned to school, pursuing an engineering and applied science doctorate at the University of New Orleans. He wrote his dissertation about the history of human interference with Lake Pontchartrain's ecosystem, beginning in 1718, when French settlers chopped down the hardwood forests of magnolia, oak, and palmetto that decorated the natural levees of the Mississippi River. Before he completed his dissertation, he was hired by his mother's old employer, the Army Corps of Engineers.

The corps was the latest institution charged with managing the environment of southern Louisiana and, Lopez suspected, no better than its predecessors. Lopez joined its coastal-restoration program, which he soon concluded was not nearly adequate to the task. The basic problem was twofold: even as land loss progressed, global warming amplified the threats posed by hurricanes. It was too late, Lopez concluded, to rely on levees, flood walls, and artificial reefs. Nor was it sufficient to preserve what remained. New land had to be added.

On nights and weekends, Lopez developed his own plan to save the Louisiana coast. His solution resembled the one that has been repeatedly reached by climate policy agonizers during the last forty years. The problem had to be attacked from every angle, in a coordinated fashion, cost be damned. Implicit in his strategy was the acknowledgment that a stable coast required constant, and profound, human intervention. Lopez believed this principle to be true not only of Louisiana but of the entire planet. "There's no such thing as a pristine environment, an environment that can be left on its own without being managed," said Lopez. "We're past that point. It's like Colin Powell said: 'You break it, you own it.' Well, we own it now." Nothing less than a maximalist approach— borne by desperation, terror, and an unshakable belief in human ingenuity—would suffice. Lopez called it the Multiple Lines of Defense Strategy.

Lopez proposed his moon shot to the corps in June 2005. It was greeted politely. His boss told him, "John, this is a really good idea. But you need to understand, the corps can't do this." Lopez agreed. The corps' bureaucracy was too balkanized to allow for the kind of systemic campaign that Lopez knew was required. In July he presented his paper at a NOAA meeting in New Orleans and was awarded a prize named for the activist Orville T. Magoon, honoring the greatest contribution to the public understanding of coastal threats. Still nobody took him seriously. Louisiana's bureaucracy was as sclerotic as the corps'; how could it possibly address a problem of this scale? And where would the impoverished state find the billions to fund such a plan? A month later, Katrina hit.

In the subsequent period of dread and opportunity, Louisiana merged its coastal-restoration and flood-control divisions, creating a centralized entity called the Coastal Protection and Restoration Authority. CPRA—"sip rah"—set about formulating a grand plan to preserve the coast. Lopez's Multiple Lines of Defense Strategy became its organizing principle. The first Comprehensive Master Plan for a Sustainable Coast was ratified in 2007. It is redrafted and reauthorized every five years. It contains one hundred and twenty-four projects designed to create tens of thousands of acres of land, preserve what remains, and build fortifications against hurricanes and sea level rise. The state did not characterize the plan as the world's most expensive climate-change-adaptation plan, but that was one way to describe it.

The master plan spoke of "coastal restoration," of "rebuilding" and "saving" the wetlands. These were euphemisms. What was lost cannot be restored, no more than the past can be relived. The best that could be achieved was an artful simulacrum of a delta that served the same ecological ends. The master plan was the blueprint for this simulacrum. In one of the reddest states in the Republic, it enjoyed thunderous bipartisan support, endorsed by scientists, the tourism bureau, the energy industry, every environmental organization in the region, Bobby Jindal, and John Barry.

Under Barry's leadership, SLFPA-E began pushing aggressively to commence the largest land-building projects, which targeted the marsh south of New Orleans. The state estimated that the master plan would cost fifty billion dollars. The state had not figured out who was going to pay for it.

John Barry nominated the oil and gas industry.

$$\oplus$$

Barry's argument was simple: The industry should pay its fair share. Eighteen billion—thirty-six percent of the bill, the percentage of damage for which the industry claimed responsibility—seemed like a good start. He did not expect ExxonMobil, Shell, and Koch Industries to offer billions to coastal restoration out of the generosity of their corporate hearts. But he believed they were legally bound to do so.

Barry's claim drew on the wording of the permits granted to oil and gas companies since the 1920s. A company was allowed to carve up the marsh as long as it repaired the damage it caused. These terms became more explicit in 1980, when Louisiana began adhering to a federal law that ordered companies to restore "as near as practicable to their original condition" any canals they dredged. (Barry, when reciting this clause, gave violent emphasis to the *tic* in "prac*tic*able.") Most companies didn't bother to refill their canals, however, and the state failed to enforce the law. This was unsurprising: until scientists began understanding the significance of the wetlands to the ecology of southern Louisiana—and to the fate of New Orleans—nobody besides a few poor fishermen took any interest in the condition of the marsh. "Swamps were the equivalent of wasteland," said John Lopez. "The terminology was almost interchangeable." Many of the projects in the master plan called for plugging canals that the companies, according to their permits, should have restored decades earlier.

Barry believed his claim was overwhelming. "You look at the photographs of the damage, and you say, what the *fuck*? Are you out of your mind? They violated the terms of their contract. They broke the law!"

The levee board was powerless to enforce the law—despite its grand mission, it was, as its president, Stephen Estopinal, was fond of saying, "an authority without any authority." It could serve only as consultant to other agencies, particularly

the Army Corps of Engineers—a body not especially known
for welcoming outside help (a tragic motif in *Rising Tide*).
Barry wondered, however, if the levee board could sue the
companies that had defaced the marsh. The complexity of
such a case would be daunting, baroque. Lawyers would have
to determine who dug every foot of the ten thousand miles
of canals and pipelines and quantify, in dollars, the extent
to which each had endangered New Orleans' flood defense.

Barry couldn't find anyone who shared his appetite for
the fight. Academic centers and environmental organizations
lost interest when they calculated how many millions of dollars would have to be spent on expert witnesses and coastal
surveys. Barry needed not only legal expertise but also the
money to take on Exxon, Shell, Chevron, and about ninety
other of the world's richest companies at the same time.

He compiled a list of lawyers who had success bringing
major environmental lawsuits in Louisiana. It wasn't much
of a list. The first name was Gladstone N. Jones III, an attorney who split his time between New Orleans and Park
Avenue. In 2006 he had won the largest judgment against
the oil industry in state history: fifty-seven million dollars
in damages from ExxonMobil for pollution and marshland
destruction. Beating Exxon in Louisiana was an accomplish-
ment on the order of beating Goliath in the Valley of Elah; no
other candidate could compete. Jones—picture a mid-career
Fred Gwynne, with the same deep, cajoling voice—didn't
think long about accepting the case. He figured it would be
"a pretty easy one to try, because the damage is so clear. I've
never had a case where the industry acknowledges damage
in their own papers. The science is irrefutable. The cause is

irrefutable. The oil companies didn't fill in the canals. Why? Because it cost money."

Jones' experts concluded that ninety-eight corporations had violated their permits. Barry hoped that the New Orleans levee board could serve as a "heat shield" to deflect the industry's furor, giving additional parishes cover to file their own tack-on lawsuits. Cover was needed; there was no precedent for an environmental lawsuit of this scope in American history. "New Orleans and the greater area will largely disappear over the next fifty years if nothing is done," said Jones at the time. "That's not lawyer talk or political talk. It's reality."

On the morning of July 24, 2013, Barry announced the lawsuit at a news conference held in the Orleans Levee District's safe house, a structure elevated twenty feet off the ground inside a warehouse built to withstand one-hundred-and-eighty-mile-an-hour winds. Barry was unnerved by the ceremony of the event. He was a veteran of book tours and network interviews, but this felt more like a weigh-in before a championship bout. He stumbled immediately, announcing that the board had filed a lawsuit against "ninety-seven oil and gas *attorneys*," when he meant to say "companies." He cursed loudly into the microphone and began again.

"From now on," one of his lawyers said afterward, "I'm going to have my interns start my car."

Everyone laughed, except Gladstone Jones. "This," he said, "is going to get dirty."

"What are they going to do?" asked Barry. "Not buy my next book?"

Barry figured there would be a period of quiescence while the industry coordinated its response. Within hours, Jindal,

who was in Aspen at a meeting of the Republican Governors
Association, released a statement. The suit, he wrote, threatened "our coast and thousands of hardworking Louisianans
who help fuel America by working in the energy industry."

The fight over the lawsuit had entered a new phase. Barry
wasn't surprised that Jindal opposed the suit. But he was surprised by what happened next.

$$\oint$$

"Louisianians who make money in oil buy politicians, or
pieces of politicians, as Kentuckians in the same happy situation buy racehorses. Oil gets into politics, and politicians,
making money in office, get into oil. The state slithers around
it." These lines, written by A. J. Liebling in 1960 at the dawn
of the offshore-drilling era, have acquired the quaintness
of folklore. Louisiana no longer slithers in oil; it drowns in
it. It is also high on natural gas, thanks to the boom in hy-
draulic fracturing. In the last century the industry has extracted more than half a trillion dollars in natural resources
from state territory. Half a trillion can buy a lot of things,
especially those that are unseen. Huey Long referred to the
industry's sphere of influence as the "invisible empire" in
1929, which was the last time that a governor could win office
by antagonizing the petroleum lobby. Soon after, the state,
which owned the oil and gas, ceded political control to the
industry, which needed the oil and gas. Environmental activists referred to Louisiana as "a petrocolonial state," and
for a century the petroimperium had been embraced locally
as a benign patriarch. The newspapers regularly published
encomiums drafted by industry lobbyists under the guise

of editorials, such as OIL AND GAS IS THE FABRIC OF YOUR TOWN, by the vice president of the Louisiana Oil and Gas Association, composed in a high liturgical register: "As residents of the oil and gas state of Louisiana, we have an obligation to pass along [our] respect for the industry that does so much for our state."

The industry's estimates of its own munificence were comically inflated; taking into account "indirect" impacts, it claimed responsibility for three hundred thousand jobs. The real figure was closer to sixty-five thousand jobs, which was significant enough—only seven percent of the state workforce, but it dwarfed every other industry. "If you want to know what the Louisiana economy might be without oil," said Andy Horowitz, a historian of disasters at Tulane, "look at Mississippi." Mississippi would be only a slight demotion. Louisiana ranked second in cancer incidence; second in infant mortality; third in poverty rate; forty-seventh in life expectancy; and forty-seventh in median income. What, then, had the industry's riches bought? The same thing they bought in Liebling's time: politicians.

Bobby Jindal had received more than one million dollars in campaign contributions from the oil and gas industry. Barry had expected that Jindal would reply in a spirit of "malevolent neglect"—that he would condemn the lawsuit while grudgingly acknowledging his constitutional inability to block it. Jindal, after all, had been under pressure to come up with funding for the master plan. But the immediacy and vehemence of his response was, as one New Orleans political consultant put it, "a dog whistle for a lot of people in the state."

There were indications that some of the defendants were

amenable to a settlement. A onetime contribution to the master plan would benefit the industry's image and resolve uncertainty for shareholders about future liability for coastal land loss. "They want to get all this stuff off their books," the state's attorney general told a local radio host, predicting "some big-time movement." But Jindal, who was already broadcasting his ambitions for the 2016 Republican presidential nomination—ambitions ridiculed nowhere so much as in his home state—made clear that he would dedicate his governorship to sparing the industry the indignity of a lawsuit. Saving southern Louisiana itself appeared not to fit his timing for running for president.

Jindal's first gambit was to erase Barry. The governor declared that he would refuse to appoint any candidates to the levee board who supported the lawsuit. Barry happened to be up for reappointment. The board had been designed explicitly to insulate its experts from political meddling; an independent committee needed to approve the governor's candidates. But the director of that committee, Jay Lapeyre, was not himself independent; he was the board chairman of a large oil and gas technology-service company.

Lapeyre began the next meeting by saying that while he was under no pressure from Jindal's office, he wanted to reconsider Barry's nomination, because the lawsuit had thrust him into "the political world." The committee rejected Barry. In his place, and for the other open seats, Jindal selected candidates who opposed the lawsuit and lacked expertise in flood protection. One was an oil company lawyer. After Jindal's appointments were confirmed, a third of the levee board opposed its own lawsuit.

Barry might have concluded that after eight years of public service it was time to get back to writing. Instead, within days of his ouster, he raised money for gasoline and a part-time assistant and set off on what resembled a statewide book tour, only instead of a book he promoted his lawsuit. In Rotary Clubs and newsrooms, in cities and small hamlets, he explained the lawsuit's rationale and showed before-and-after photographs of the marsh. He believed that he was making a difference.

When the Louisiana legislature entered session, Jindal announced a new tactic: the passage of a bill to kill the lawsuit before it came to court. The new law would have to operate retroactively, going back in time to forbid the levee board to sue in the first place.

The "legislature was a swamp, 'timid and third-rate,' thick with petty men greedy for small things," wrote Barry in *Rising Tide*. He was writing about Mississippi in 1910 but might as well have been describing the situation in Louisiana more than a century later. During the term, seventy oil and gas lobbyists flocked to the state capitol, about one for every two legislators. "When the industry feels threatened," said Darrell Hunt, one of the few lobbyists representing the levee board, "it's like a military base being under attack."

Still Barry felt optimistic. Legislators didn't have to champion his lawsuit; they only had to agree to allow the courts to decide the case. He had already survived a smear campaign: he was called an "environmental extremist" ("The worst smear they can think of in Louisiana") and an "aspiring author." Virtually every editorial board in the state condemned the

effort to kill the lawsuit, including those in the industry don-
jons of Lake Charles and Houma. Three former governors of
Louisiana supported the lawsuit, as did three of four southern Louisianans.

The ferocity of the lobbying effort roused the suspicion
of some legislators. "If the industry doesn't believe the levee
authority has any standing to file a lawsuit, they should go
to court and file a motion," said John Bel Edwards, a Democratic representative from Roseland who was considering a
long-shot bid to succeed Jindal as governor. "Rather than go
to the courts, they ran to the legislature. That was the first
indication that the levee board's claims were not frivolous
at all, but had merit." Daniel Martiny, a Republican senator
from the New Orleans suburbs, put it more directly: "The
bottom line is the oil industry has been a big supporter of
a lot of people, myself included. At times, I was even one of
their heroes. And I had people way up in the oil-industry
hierarchy tell me, 'We know you're right about this, but we
don't have any other way to stop this.'"

One by one, bills written to kill the lawsuit failed—more
than a dozen in all. With a week left before the legislature
disbanded, one of the levee board's lobbyists told Barry, "If
we voted today, we'd win."

Louisiana grants its governor more powers than nearly
any other state in the nation. During the final days of the
session, Barry watched helplessly as legislators began mys-
teriously to change their minds. "You could feel the weight
coming down," he said. "You could see them getting peeled off,
one at a time." He heard that Representative Gene Reynolds, a
Democrat from Minden, was told that funding for a new roof
for a VFW hall in his district would be eliminated if he didn't

support a bill against the lawsuit. (Reynolds denied this, sort of, saying that his vote "was based on information from my district, which depends heavily on oil and gas and the related benefits.") Senator David Heitmeier, a Democrat from Algiers who was an optometrist, had supported the lawsuit, only to reverse his vote at the last minute. A week later, Jindal signed a bill allowing optometrists to perform some minor eye surgeries—a pet cause of Heitmeier's. (After leaving office in 2015, Heitmeier returned to his optometry practice, which advertised its ocular surgery services.)

Many legislators didn't need to be persuaded. Robert Adley, a Republican from Benton who was one of the senate's leading power brokers, called it "absurd to say that the oil and gas industry has damaged the coast. They did what they were told to do." This was a claim that most industry representatives had not been brazen enough to make themselves. Then again, Adley himself was an industry representative, the owner of the Pelican Gas Management Company for the previous thirty years, and for twenty years before that president of ABCO Petroleum.

State legislator was a part-time job and, for many Louisiana legislators, a second job. Representative Gordon Dove, a Republican from Houma who was chairman of the Committee on Natural Resources and Environment, owned Vacco Marine, which provided vacuum trucks for the cleaning of oil tanks (when it was not busy contesting complaints from environmental regulators). Representative Neil Abramson, a Democrat from New Orleans, was a lawyer who defended oil and gas companies in environmental-damages lawsuits. Representative Jerome "Dee" Richard, an independent from Thibodaux, was a sales manager for a Chevron contractor.

Representative James Morris, a Republican, was an independent oil producer from Oil City.

But many of the same legislators had watched for years as their districts had sunk into the Gulf of Mexico. Those who supported the lawsuit had emotional reasons for doing so. "I have hunted and fished in the marsh for more than fifty years," said Senator Conrad Appel, another Republican from the New Orleans suburbs. "I've seen the whole outer edge of the marsh back up, year after year, before my very eyes. Even vast stretches of land that were there twelve months ago are now gone. And it's accelerating dramatically. Once it starts, it just *goes*."

Dee Richard's district included Lafourche Parish, which had suffered some of the most extreme land loss in the state, with an especially high proportion of that loss attributable to oil and gas canals. Richard didn't like the idea of a bill that would kill a lawsuit already in progress and pledged his support to Barry. But everyone else in his delegation opposed the lawsuit, so he went along. "I didn't want to be the only one voting against it," he admitted.

Once it became clear that a bill would pass, some peculiar votes—or nonvotes—followed. Abramson, the oil and gas lawyer from New Orleans, declined to vote. Senator J. P. Morrell, also a Democratic lawyer from New Orleans, voted for the bill and later claimed that he had not been present for the vote and that his machine had malfunctioned. Senator Gregory Tarver, a Democrat from Shreveport, claimed to have voted for the bill by accident. No one who voted against the bill claimed it was by accident. "What we enacted was horrible public policy," said Representative John Bel Edwards, "and it was because of the pressure put on by the oil and gas industry and their paid lobbyists, and the governor trying to score points with the Koch brothers, and with Republicans around the country."

Barry accepted the defeat with his usual admixture of vitriolic rage and defiant optimism. There was nothing left for him to do except write about the lawsuit, which he had forsworn—mainly because his opponents accused him of having ulterior motives for joining the levee board in the first place. He had forced himself not to take notes about his experiences to avoid the temptation. Then he reconsidered. The book would be a kind of sequel to *Rising Tide*, about "the interplay between the river, the sea, politics, and oil over the last century." It would culminate with the battle to make the industry pay for its share of the master plan. In the climactic chapter, a figure named John Barry would appear on the scene. But John Barry, John Barry insisted, would be just a minor character.

Still the story, like the disappearance of the coast, showed no sign of ending. "The idea of making the industry live up to its legal responsibility is not going to die," Barry had said, and he was right. After a series of negative rulings regarding jurisdiction, the levee board's lawsuit failed. But after Jindal left office, he was succeeded—through the legerdemain of Louisiana general election law and an internecine feud within the state's Republican Party—by John Bel Edwards, the obscure Democratic representative from Roseland. As governor, Edwards pressured the state's ten coastal parishes to sue the oil and gas industry; if they didn't, he declared, the state would. Those ten parishes, as well as two others just inland, filed forty-two lawsuits. Gladstone Jones represented Orleans Parish. In 2019, the first of the ninety-eight defendants, a

mining company called Freeport-McMoRan, agreed to settle for nearly one hundred million dollars. The corporations
responsible for a far greater portion of the damage to the
marshes—Chevron, ExxonMobil, Shell, ConocoPhillips,
and BP, which by then had contributed four billion dollars
to the master plan as part of its settlement for the Deepwater Horizon oil spill—showed no inclination to concede.
Upon returning from a COVID-19-imposed break in the
spring of 2020, the state legislature elected, as a top priority,
to kill the lawsuits. The author of the bill that passed the
senate was Michael "Big Mike" Fesi Sr., a Houma Republican. Before running for office—his campaign slogan was
"Big Mike Cares"—Fesi worked for thirty years in oil and
gas. He was the founder and chairman of a pipeline construction company.

The session ended before the law could pass the house,
but Fesi vowed they would try again, and again, and again.

"I'm here," said Big Mike, "to save our coast."

II. Barataria

In response to complaints some years ago about blocked
plumbing along New Orleans' Claiborne Avenue, city workers opened the sewer main and found a human nose. Following the line down the avenue, popping open manholes
and looking inside, they discovered ears, fingers, fingernails,
shriveled flaps of skin, viscera. Where had it all come from?
Had a serial killer taken up residence on Claiborne Avenue?

To solve the mystery, the Sewerage and Water Board
turned to Warren Lawrence, a former plumber who served as

the utility's inspector. Lawrence conducted his job with the perspicacity of a forensics scientist. It wasn't enough for him to repair a drainage problem; he made a point of pursuing each disturbance to its source and holding the perpetrator responsible. When Lawrence encountered a section of corroded pipe, he traced the damage to a battery factory near the Superdome that had been illegally pouring acid down the drain. After finding a black-and-white jumpsuit in a sewer, he learned that inmates of Orleans Parish Prison had been stuffing their uniforms into the toilets to back up the jail's plumbing system. To increase their odds of success, the prisoners flushed at the same time. They called it a "Royal Flush."

Lawrence followed the trail of body parts to Charity Hospital. The hospital's sewer line was clogged with flesh. Lawrence asked hospital administrators why they were dumping bodies down the drain. They explained that until recently they had incinerated unclaimed corpses. The stench was abhorrent, however, so they had installed a million-dollar, fifteen-horsepower grinder pump. The machine ground the bodies into a slurry, but small parts escaped the blades. Lawrence ordered the hospital to remove the grinder. Because he was backed by the force of city hall, the hospital had no choice but to comply.

Lawrence was reminded of all those severed noses, ears, and fingers three decades later when he noticed that the home he had built for his retirement, thirty minutes south of New Orleans, was regularly being coated by fine black dust. It was no small inconvenience, the dust, because the house was gigantic, with three floors, three porches, a swimming pool, and a second-floor deck the size of a helipad. Every surface—the railings, the roof, even the bottom of his pool—was painted

white. In the driveway sat his white Toyota. At the end of his
dock floated a twenty-four-foot party barge, also white, with
white leather benches.

Lawrence was seventy-four years old. The house, the car,
and the boat represented the sum of his life's work. Though
Lawrence had ultimately risen to an executive position at the
Sewerage and Water Board, he and his wife, Gayle, had re-
mained in his cramped two-bedroom for forty-three years.
"When people came over," said Lawrence, "they would say,
'You live in this?'" But he was saving for his retirement.

After Lawrence spent months touring properties across
the gulf, one of his sons called with a lead. A luxury development was being built in Plaquemines Parish, which is ten
miles south of New Orleans but feels, to most New Orleanians, like its own sovereign island nation. A firm had built
levees around hundreds of acres of marsh and spent a decade
draining the land. They named the new town Myrtle Grove.
Homeowners would have immediate access to the abundant
fishing, boating, and natural beauty that gave Plaquemines
its nickname, Sportsman's Paradise. Lawrence drove to the
site every day to watch the crew pave the streets. The morning the lots went on sale, he was second in line. He bought
one of the largest sites, situated at the junction of two canals.

Lawrence asked his wife to design the house. She drew
a four-thousand-square-foot redoubt, with balconies cantile-
vered over the adjacent bayou. "I drew it so big," said Gayle,
"because I didn't think he'd build it." He built it.

"When Momma's happy," said Lawrence, "everybody's
happy."

Gayle wanted the house white. Lawrence painted it white.
Overnight it turned black. It took the Lawrences a day

and a half to wipe down every surface with damp rags. The house would remain white for a day, or a week, but inevitably the dust returned. Lawrence began to keep a weather log. He found that his house was turning black on days when there was an easterly wind. He walked to his rooftop deck and looked east. There, half a mile away, loomed a cordillera of black hills. They ranged across a one-hundred-and-fifty-acre coal storage terminal on the west bank of the Mississippi. The black hills were coal and petroleum coke, a particularly ghastly by-product of oil refining.

Lawrence drove to the terminal. The plant manager was not as receptive to his complaints as the Charity Hospital administrators had been. There was another coal terminal across the river, said the manager. How did Lawrence know where his dust came from? Lawrence explained that when he stood on his deck on windy days, he could see the black veils blowing off the terminal's property and advancing toward his house like storm clouds. The manager demanded proof.

Lawrence wrote letters to local politicians, the Louisiana Department of Environmental Quality, and the EPA. State officials told him that having known about the coal terminal when he bought the property, he had no standing to complain. Lawrence replied that he did not know it would regularly cough soot onto his house. Officials swabbed the white railings and ran tests and assured him that the coal dust was not a health hazard. "I don't think I can believe you," replied Lawrence. "If I can see it, how can it be doing my lungs any good to breathe it?" He thought of his father, a pipe fitter who brought asbestos home in sacks like bulk bags of flour. Lawrence and his siblings mixed the asbestos with cement to make flooring for their house. His father died of asbestosis at the age of seventy-one. His friends' fathers, who had worked at the same firm, all died within two years.

Warren Lawrence embarked on a second career. He attended every Plaquemines Parish council session, organized meetings in churches and living rooms and researched the handful of property owners, lawyers, businessmen, and politicians who controlled the parish. He inspected the parish's sewage, in other words. He dug up the bodies. And he learned that the toxic coal dust blowing over his property was just the beginning of his trouble.

At the heart of Louisiana's Comprehensive Master Plan for a Sustainable Coast was an act of defiance, overturning the fundamental principle guiding life in the Mississippi River delta for centuries. As long as European settlers had established cities and farms along the river, they had tried to control its flow and limit its eruptions. They managed to do so with reasonable success until the floods of 1927 forced the passage of a law that was, as John Barry wrote, "the most comprehensive and expensive piece of legislation Congress had ever considered." It gave the Army Corps of Engineers control of the system of dams, levees, and floodgates that pinned the Mississippi in place, preventing it from wagging back and forth across southern Louisiana like the tail of a delighted dog. The corps had fought continuously since then to achieve one of the greatest feats of modern engineering: an obedient, predictable river. Were it not for the corps, southern Louisiana would be unsuitable for modern civilization. But it was the wildness of the river that had built and sustained southern Louisiana. The Mississippi is the nation's sewer main, the repository for the Red, the Missouri, the Kaskaskia, the Des Moines, the White, the Rock, and hundreds of other tributaries that, like so many drain lines, collect particles of soil and rock and sand traveling from as far away as Potter County, Pennsylvania; the Blackfeet Indian Reservation in northwestern Montana; the Smoky Hills of northern Kansas—some five hundred million tons of sediment each year. For millennia, whenever a breach opened in the riverbank, muddy water rushed through, depositing alluvium that solidified into land. When one crevasse plugged with mud, the river opened breaches elsewhere. Since the Mississippi has been hemmed in, most of its sediment, instead of replenishing the wetlands, discharged into the Gulf of Mexico and disappeared off the continental shelf.

The master plan's major land-building projects intended to reverse this process. They were designed for Plaquemines Parish, which in the last century has withered to almost half its original size, its dwindling population clinging to the high ground on either bank of the river. Should nothing be done, Plaquemines would lose more than half of its remaining land, and one of the world's most productive ecosystems, in the next fifty years.

The master plan proposed to cut open the federal levee at two locations in lower Plaquemines, creating powerful new distributaries of the Mississippi River. When running at full capacity, the diversions would themselves rank among the nation's largest rivers, flowing at more than twice the volume of the Colorado. Over the course of years and decades, it was hoped, the gargantuan volume of sediment borne by the man-made rivers would patch the holes in the marsh's moth-eaten fabric.

You did not have to be a scientist to identify on a map of southern Louisiana the site where a diversion was most desperately needed. It was the spot, thirty river miles south of New Orleans, where the river curved east around a section of marsh that resembled tissue paper floating in a toilet bowl. Below those wetlands lies Barataria Bay, named after a village in *Don Quixote* governed by Sancho Panza, who is tricked into believing it is an island. That Barataria was not an island, and this Barataria, left to its own devices, will soon cease to be a bay, absorbed by the Gulf of Mexico.

The name of the location chosen for the Mid-Barataria Sediment Diversion was St. Rosalie Bend. A diversion placed at this turn of the river, where the level of sediment was particularly high, was expected to restore as much as fifty square miles of land in fifty years—so long as nothing got in its way.

⊕

The man-made monster of a river created by the Mid-Barataria Sediment Diversion would surge through a plot of land that was once the site of the St. Rosalie sugar plantation. St. Rosalie was established in 1828 by a free man of color named Andrew Durnford, a strict slave master who struggled to make a profit, despite the forced labor of seventy-five men and women. He died indebted to the white slave owner who lent him the money to purchase the land; his heirs were forced to sell the property at a loss after the Civil War. Between St. Rosalie Plantation and Myrtle Grove—formerly the Myrtle Grove Plantation—lies the hamlet of Ironton, which occupies the former Ironton Plantation. After emancipation, Ironton's slaves gained ownership of the land, and their descendants remain there today, five generations later. But each generation has had to fight to stay.

Ironton was best known for the abuse it suffered from Leander Perez, the rabid segregationist and anti-Semite who served as parish dictator from 1919 until his death in 1969. Perez presided over one of the most powerful political machines in southern history and, in the process, siphoned eighty million in oil royalties into his personal bank accounts. Hailed by his own church as "the leading racist of the South," Perez neglected to authorize levees for the predominantly Black areas of his parish, wrote laws that prevented Blacks from voting, built a prison camp for freedom riders, and blocked desegregation in Plaquemines schools, at one point personally interrupting a catechism class at a local church to expel a Black student at gunpoint. Black Plaquemines residents recalled grandparents who refused to speak Perez's name louder than a whisper in their homes. After his death, his sons, who had inherited his flair for inventive racial cruelty, inherited his power. As late as 1980 the parish had segregated hospital waiting rooms, public parks, and hurricane evacuation plans: whites could take shelter at a local school, while Blacks had to seek refuge at a navy station seventeen miles away. Blacks were barred from government jobs, and the parish council declined to apply for federal antipoverty funds. But Ironton had it worst of all. Because the council refused to provide running water or a sewer system, residents were forced to collect water in cisterns, as the settlers of New Orleans had done two centuries earlier. It was not until 1981, after exposés by Walter Isaacson in *Time* and

Dan Rather for *60 Minutes*—and an attritional power struggle between Perez's sons—that the parish granted Ironton
water.

The relationship between the parish's Black population and its white leadership has not much improved since.
Burghart Turner, Ironton's council representative, accused
the parish of delaying initiatives in his district and failing
to conduct basic maintenance and repairs. Though the energy plants boasted of hiring locally, many defined "local"
as within a five-hundred-mile radius, meaning some local
employees lived in Memphis and Tallahassee. Turner questioned the motives behind a diversion that would carve a
barrier between the wealthier, predominantly white areas of
northern Plaquemines and his district, which contained the
majority of the parish's Black population. "You're physically
cutting the parish in half," said Turner. "When you do that,
you make it easier to lose the entire landmass." For generations the Perezes and their constituents had tried to cut the
parish in half. But now the partition would be manifested on
the map.

Warren Lawrence's dream house also sat on the southern side of the planned diversion. Ironton and Myrtle Grove
had roughly the same populations but inverse demograph-
ics: the former was entirely Black and poor; the latter was
entirely white, with most of the properties serving as vacation homes. Before the announcement of the Mid-Barataria
Sediment Diversion, said Audrey Trufant-Salvant, Turner's
administrative assistant, "nobody from Myrtle Grove ever
talked to anybody from Ironton." That changed when Lawrence started asking questions about the diversion. He was
amazed by what he discovered. The Mid-Barataria diversion,

the crown jewel of Louisiana's master plan, had come under threat—from the State of Louisiana. The state had granted a coal company named RAM Terminals, an affiliate of Kentucky's Armstrong Coal, permission to build a major storage facility on the exact plot of land marked for the diversion.

"I have concerns about the diversion," said Lawrence. "But the coal terminal would destroy us once and for all."

When Lawrence met with residents of Ironton, he was surprised to learn that they suffered from coal dust pollution too. It was a constant irritant in Ironton, responsible for a variety of chronic health conditions. But the grain dust was worse. A nearby grain elevator shed particles that fell over the town like yellow snow every time there was a westerly wind. The east wind brought coal; the west wind brought grain—residents could tell which way the wind was blowing by the color of their porches.

"There isn't a kid born in Ironton without respiratory problems," said Trufant-Salvant, whose great-great-grandfather, a member of the family who owned Ironton Plantation, is buried in the town's cemetery. She kept her home spare and neat because she had to clean it four times a week. On some days dust swirled on her porch like a cloud of midges. "This is the country. We're used to keeping our doors open, our windows open. But now we keep our doors closed all the time."

Trufant-Salvant wore pearl earrings and a navy blazer and carried herself with the bearing of a stateswoman. Her even-tempered tone was at jarring odds with the abuses she related. She had first heard the term "industrial corridor" a year earlier, but it had quickly become the default qualifier in articles about Ironton. Billy Nungesser, the parish president, used the term liberally and applied it retroactively, as if this section of the west bank had long ago been granted a formal designation. "Anybody who lives there," he said, "knew that corridor along the river in Plaquemines Parish was going to be industrial." Trufant-Salvant knew no such thing.

During Nungesser's term, many of the citrus farms that used to line the lower Mississippi were sold to heavy industry—not just the three coal terminals, the grain elevator, and the oil storage facility, but also the Conoco-Phillips Alliance Refinery and the Plaquemines Port complex, which served petroleum and coal tankers. These all lay within a six-mile span encompassing Myrtle Grove and Ironton. Trufant-Salvant remembered when Ironton children swam and fished in the Mississippi River. Nobody went near the river anymore. It was the second-dirtiest river in the United States (after the Ohio, its largest tributary), and nowhere was it dirtier than when flowing past Ironton.

RAM had no objection to allowing the diversion to run through its property, the new river coursing alongside eighty-foot-tall piles of coal. An independent study found that the terminal would reduce the amount of sediment carried into the marshes by as much as half a million tons over ten years, while pumping coal runoff—toxins like arsenic, polycyclic aromatic hydrocarbons, and sulfides—into the wetlands, poisoning the very flora that the diversion intended to save.

Although the director of Louisiana's coastal authority, Jerome Zeringue, expressed "serious concerns," the state decided that the reduction of half a million tons of sediment was not "inconsistent" with the aims of the master plan, and granted RAM's permit. "I can understand that it doesn't send the right message," said Zeringue. "We would prefer not to have it there. The question is, can we stop it?" Zeringue, the state representative, didn't think the state could. "It'd be great if we could lock up the whole river for the purposes of coastal restoration. But obviously we can't do that."

The people of Ironton had been hearing the word "obviously" for five generations. "Anybody who is willing to issue a permit to RAM Terminals, knowing that the coal is going to have an adverse effect on the diversion, can't be concerned about coastal restoration," said Burghart Turner. "If we're serious, then why are we allowing more coal to pollute the marshes? You don't play around with cancer. You remove it. As the saying goes, 'What you do speaks so loudly that I can't hear what you're saying.' Go sell that story to someone else. Let me put my dunce hat in the closet, that way you quit talking to me as though I'm stupid, thinking you can sell me anything."

All roads in Plaquemines Parish led to—and most of them had been rebuilt by—Billy Nungesser, whose constituents addressed him as "President Nungesser." Leander Perez had been known as "King." Out-of-staters might most easily picture Nungesser as a swampland Chris Christie. He shared the former New Jersey governor's insouciant charisma, undulating build, and winsome indifference to the art of self-censorship that most politicians master by their first pancake

breakfast. Two years before Christie gained national sympathy for his indignant response to Hurricane Sandy, Nungesser emerged in the aftermath of the BP oil spill as the
state's refractory, brokenhearted spokesman, serving—at
least on cable news—as a foil to Bobby Jindal's mugged-
schoolboy pose. Sunburned and greasy from a day spent
riding airboats through crude-slicked wetlands, tented in
billowing blue oxford shirts (necktie long since cast aside),
and hitching up white shrimp boots that he called his "Ca-
jun Reeboks," Nungesser demanded that Washington bu-
reaucrats be strung up for their incompetence. After BP's
chief executive visited Plaquemines and denied the presence
of large oil plumes in the gulf, Nungesser told the executive
he was "lucky he got out of here alive." He appeared to take
the BP disaster personally. When President Obama toured
the coast, Nungesser traveled to Grand Isle to confront him:
"There's been a failure of leadership at all levels. Who in the
hell is in charge?" It was clear, in Plaquemines, who the hell
was in charge.

The four billion dollars received in the settlement from the
BP oil spill lawsuits provided a convenient down payment on
the master plan, enough to begin work on the Mid-Barataria
Sediment Diversion and the second major diversion, on the
east bank ten miles south, the Mid-Breton Sediment Diversion. Nungesser could take some credit, then, for helping
to bring the master plan to life—which he did, often. Later
he'd claim that his efforts to develop the master plan were
responsible for winning him reelection as parish president
with more than seventy percent of the vote.

Nungesser's throne was in a corner office in the glass-
wrapped parish headquarters, which was raised on stilts ten

feet off the ground in Belle Chasse. Ronald Reagan, astride a white horse, peered down from the wall; in a second photo, Reagan shook a young Nungesser's hand. (Nungesser's father had been the state's leading Republican kingmaker.) Three flags stood behind the desk: the American flag, the state flag, and the flag of Plaquemines, a green triangle with bars of red, yellow, and blue, framed by a white border. The green triangle represented the growth of the delta. The whiteness of the border represented purity.

"I really think there's a happy medium on everything," said Nungesser. "I believe in business and industry. I love animals. I love the pelicans. I love the barrier islands." His tone shifted, however, when asked whether he had neglected Councilman Turner's district. "After Katrina, I bought ten thousand dollars of washers and dryers and put them all over our parish. I rescued thirty people from that community who lived with me for a month. I paid electrical bills for people, far more than I could afford, because I love helping people." He mentioned the thousand-dollar personal checks he wrote to every parish church, the horse-riding camp he started for disabled children, the check he sent to Mothers Against Drunk Driving after he was pulled over for driving drunk as a young man. "Listen, I'm a conservative Republican. I got beat up when I gave a lot of money to people whose political ways I don't believe in. I did it to get their ear—to ask them to do the right thing. I went to Washington, week after week. Does that sound like a guy who doesn't care about the people of Ironton?"

Nungesser lived on an estate built on a man-made hill in front of a man-made lake just south of Myrtle Grove and the coal terminal. Did he have problems with coal dust?

"I get a film on my car," he said. "Is that coal dust? I don't know. I don't identify it as coal dust. I don't know if that's pollution in the air, dust, or whatever."

Then he was back on the subject of his "enemies" for another hour.

"Shame on them," he said. "They're going to have to answer to God one day. Mr. Turner calls himself a holy man? I call him the devil."

Nungesser and his antagonists were not as far apart as it might have seemed. Plaquemines Parish had made an unholy bargain with the energy industry. But nobody makes such a bargain without a reason; you do so only if you believe you've been forsaken. And the parish had been forsaken after Katrina. The White House's plan to rebuild the levee system excluded the lower part of Plaquemines; a Bush administration representative questioned whether including it would be "economically justifiable." The unsubtle implication of this statement, and others like it in the months that followed, was that Washington was happy to sacrifice southern Louisiana to the rising seas, amputating it like a gangrenous digit.

Some local activists harbored a counterintuitive theory: Nungesser was luring heavy industry to Plaquemines to extort the federal government into paying for flood protection. Begging for money to protect poor people in a flood zone would not pass congressional approval. The parish needed hostages. The more industry it could attract, the more effective the blackmail note.

Nungesser accepted this interpretation. "If I had my choice, would I choose this kind of industry?" he asked. "I don't know. But I do know that we're going to have another hurricane. With industry there, we have a fighting chance to save these communities."

Nungesser's gambit would be vindicated. After the rapid buildup of industrial plants along the lower Mississippi during his administration, the federal government had a change of heart. In 2012, the Army Corps of Engineers announced a $1.4 billion plan to improve the levees between New Orleans and the southern tip of Plaquemines, extending their length and raising them to four times their previous height.

Those levees will be worthless if the marsh continues to erode and the diversions are never built. But if enough energy facilities moved into harm's way, the strategy went, the federal government might agree to cover the entire master plan. Then it would be Nungesser, who had ascended to lieutenant governor and was expected to run for governor after John Bel Edwards left office, whose portrait would be hung in the Plaquemines government building, riding a white horse.

After two years spent gathering signatures, commissioning scientific studies, and consulting with lawyers, Lawrence and the residents of Myrtle Grove won a class-action suit against the coal terminal that had blackened his house. The judge gave the facility, which had renamed itself International Marine Terminals, four years to install sprinkler systems called

Rain Birds, cannons that spray water to wet the coal and keep it from floating away. (Other states required Rain Birds by law.) After four years, IMT finally installed the Rain Birds, but Lawrence couldn't tell if they made a difference. By then IMT had completed a nearly two-hundred-million-dollar expansion, doubling its capacity from ten to twenty million tons of coal.

Lawrence had better success against RAM Terminals. After six years of battles over environmental permits, and attendant financial losses, RAM sold the property. It was bought by a consortium joined by the Plaquemines port authority, a semiautonomous public-private partnership. The anonymous buyers intended to use the land as another shipping terminal—this time for the export of crude oil. Oil posed about the same threat to the diversion as coal; only the political interests, the revenue streams, and the mechanisms of legal liability were better camouflaged by a palimpsest of shell companies. The permit postponements continued. Still Warren Lawrence figured that the day would come when Plaquemines Parish would have to choose between land and oil—or drown in water.

III. The Forest Machine

While the parish governments and the oil and gas industry fought over who would pay to rebuild the coast, and the state fought with itself over environmental permits, a surprising new front opened in the war for Louisiana's future. A group of Plaquemines residents charged into the fight—*against* the sediment diversions that meant to save them. They pointed out that the master plan would not serve everybody. It would require the government to seize private land and render the coast unrecognizable—or rather, to distinguish from the coast's current anomalous appearance, unrecognizable in new ways. The plan would do more than test the limits of our capacity to engineer our environment. It would also test the government's ability to compel even a small, relatively powerless group of people, against their will, to suffer in the name of climate policy. The master plan would benefit the many. It would also harm the few.

Kindra Arnesen spoke for the few. She grew up in Buras, near the mouth of the Mississippi River. Her middle school was a plot of marsh a hundred yards off the coast. At twelve, after her mother lost her job, Arnesen began skipping school to walk to the harbor. A dredge boat ferried her to Bay Adams, where she met a crew of oystermen. They gave her a flatboat, rubber boots, burlap sacks, and a hatchet. With a rope looped around her waist, she trudged through the marsh, navigating the mud banks and the tufts of saw grass, tugging the boat behind her. The word she used to describe the scene, three decades later, was "serenity."

It wasn't hard to find oysters then; they were everywhere. She bent into the water—dunking her head when necessary— yanked out a cluster, shook off the mud, tossed it in the boat. When the boat was full, she climbed onto it. She cleaned the oysters, hacking off debris and dead shells, and fed them into the sacks. By the end of the day she'd have filled ten, earning a hundred dollars, the cash in an envelope that she picked off the hood of the foreman's truck. She supported her entire family, with enough left over for Girbaud jeans, Z. Cavaricci high-waisted pants, and white K-Swiss Classics.

After oyster season she fished for mullet, shoveled ice at Wet Willie's, or worked the deck on shrimp boats that left after dark and returned at dawn. The summer she turned fourteen, she and a girlfriend unloaded hundred-pound sacks for aging Vietnamese oystermen. The girls hauled as many as eight hundred a day, for a dollar a sack. "As a young girl in a port town," she later said, "a lot of bad stuff could've happened to me. Instead of getting in trouble, I worked on an oyster boat. The men and women I worked with taught me to stick up for myself. They saved me from the big bad world."

Arnesen had since devoted herself to protecting those same fishermen from the big bad world. She ran her own fishing business, bringing amberjack, pompano, black drum, and shrimp to restaurants in New Orleans and distributors that shipped up the East Coast. Most days, when not at sea, she drove between Venice, the last town before the Mississippi emptied into the gulf, and New Orleans, about ninety minutes north, buying boat parts, signing paperwork, and unloading thousands of pounds of fish from her Chevrolet Silverado 3500 pickup. She had nevertheless found the time to become one of the most prominent national advocates for gulf fishermen. Since the BP oil spill, she had attended just about every public meeting or legislative session concerning the future of the Louisiana fishing industry, which provided a third of all seafood caught in the continental United States. Like John Barry, John Lopez, Audrey Trufant-Salvant, and Billy Nungesser, Arnesen spoke of a sense of obligation: "If we don't fight for these fishing families, if we lose a couple of links out of the generational chain, we lose a whole way of life in this country." It was a fight to the death.

That was how Arnesen found herself on an equatorial afternoon at Belle Chasse Auditorium, seventeen river miles south of New Orleans, where she had come to confront the architects of the master plan. Belatedly, only after it became clear that popular support was weakest in the part of the state where the largest projects would be built, the state's coastal authority had launched a monthly outreach series called Coastal Connections. CPRA had no dedicated public relations officers, so it dispatched its lead engineers, who did not appear to relish this part of their job. The engineers were not what you'd call people persons. They were integrated-compartment-model and hydrologic-calibration persons: Brad Barth ("I've had no public speaking training"); Rudy Simoneaux ("as engineers, by nature, we'd rather be talking about geotechnical designs"); and Dain Gillen ("I'm an introvert; honestly, I'd rather be at my desk"). After mounting posters on display boards and straightening stacks of pamphlets, they summoned expressions of friendly forbearance, like science fair entrants bracing to be questioned by the judges. They were native Louisianans, with sugarcane accents and retiring demeanors, dressed in innocuous bureaucrat-casual (oxfords, slacks, brown loafers). But as the engineers shifted their weight at the far end of the vast blond parquet, they had the vulnerable air of trespassers in a foreign country where they were not trusted, did not belong, and were not wanted.

They greeted Arnesen with wary decorousness. "We have to be present or they can say there's 'no opposition,'" she said. "I see this as doomsday. This will end us."

This was the plan to send man-made rivers crashing through Plaquemines Parish. Besides enabling the parish to remain habitable in future decades, the diversions would invite back lost species and increase the biological diversity of the wetlands. In the short term, however, they would transform the delicate estuarine ecosystems. Plaquemines had the largest commercial fishing fleet in the continental United States. Arnesen was certain that the diversions would destroy it.

Arnesen was careful to emphasize that she worried about the dying marsh loss as much as anyone else: "Nobody wants land more than we do." But the diversions worried her a lot more. They would turn brackish water fresh, killing saltwater vegetation and ravaging the local fisheries. They would massacre giant populations of oysters, brown shrimp, blue crab, and dozens of species of fish. Speckled trout, to take one delicious example, will spawn only on a full summer moon, when the water is warmer than seventy-three degrees Fahrenheit and salinity levels are higher than fifteen parts per thousand. Flooding the wetlands in the late spring, when the diversions were expected to run at maximum capacity, would each year annihilate an entire generation.

The engineers' responses to Arnesen's concerns were pallid, technocratic. They pointed out that the presence of brackish water so close to the river was a historical anomaly; some of the most productive fishing areas, not long ago, had been solid land. They noted that if they failed to build land, not only the fisheries but the parish itself would, in the coming decades, vanish entirely. And they argued that the diversions would bring new species to fish, perhaps one day even making the fisheries more productive.

The high-handed dismissiveness of the engineers gave Arnesen fits. Fishing another species was not as simple as changing your lure. Boats weren't cheap. A small skiff could cost thirty thousand dollars; a larger shrimp trawler topped seven hundred and fifty thousand dollars, without accounting for gear and licenses. Most fishermen could not afford to diversify or wouldn't know how. A shrimper would lose his house if he had to run a catfish business; the tourists who traveled from all over the world to hire speckled trout charter captains would not travel for shad. Oystermen were in the most precarious position of all. Oyster leases, rented from the state by the acre, had terms of fifteen years. It was about as easy for an oyster farmer to start fishing largemouth bass as it would be for an alfalfa farmer to raise pigs. It was possible but difficult, risky, and usually cost prohibitive.

The master plan aimed to "balance" a suite of objectives: to "provide flood protection, use natural processes, provide habitat for commercial and recreational activities, sustain our unique cultural heritage, and support our working coast." This was unimpeachable in theory, offering something to everyone. But the plan was silent on what to do when these objectives came into conflict. What happened when "flood protection" threatened "unique cultural heritage"? Or when "natural processes" interfered with a "working coast"? It drove Arnesen crazy, the refusal to acknowledge that not all objectives were treated equally. It was obvious to her that the state cared more about the oil and gas industry than the fisheries, and that it worried a hell of a lot more about keeping

New Orleans dry than Buras. Most maddening of all was the engineers' emphasis on the future over the present.

"They call us resilient," she said. "We're sick of the word. Yeah, we're resilient, but we're not bulletproof. We can't keep coming back from disaster. Superman is a fictitious character."

This kind of talk disturbed the engineers. Dain Gillen, who had the least public speaking experience among them, took it personally. "It really does kind of suck," he said, on the high school parquet. "We're trying to do good. But to see people upset, or furious, getting in your face . . ."

Arnesen responded by recounting stories about the many fishermen she knew who feared that their fragile livelihoods would soon be unrecoverable. What's a man good for, if he can't support his family? a friend had recently asked her, two days before hanging himself.

"I don't just do this because it's my living," said Arnesen as she left the auditorium, trailing an entourage of well-wishers. "They've made our community feel like we're the trade-off and we don't matter. It's easy for the state to say they're going to come up with an adaptation plan. But what's the point of an adaptation plan if the end goal isn't the survival of the people you're trying to save?"

The survival of the people. How to characterize the way of life threatened by the diversions? It did not simply entail the right to fish the same species that your grandfather fished, or to inhabit the same arpent, or to live off the fruit of the sea and land, though that was all part of it. It would be laughably inaccurate to say that the lowest stretch of the Mississippi was more remote than any other rural area of America, though often it felt that way. To an interloper from New Orleans, the lower Mississippi looked like the end of the world, a wilderness untouched by human interference—even though the land owed its existence to human interference. You could say, at the very least, that it remained possible there to live a life of wildness and freedom. This was especially true for those living on the wrong side of the Wall.

It was not technically a wall, but that's how it was known in Plaquemines. Officially it was the Hurricane and Storm Damage Risk Reduction System, or HSDRRS ("his dress"). The Army Corps of Engineers constructed its $14.5 billion network of interlocking gates, levees, and flood walls in response to the levee failures after Hurricane Katrina. It was designed with the explicit goal of protecting the New Orleans metropolitan area from a catastrophic hurricane. To do so, the corps had to draw a line separating those who would be kept safe from those who would be abandoned to the furies. The entire east bank of Plaquemines Parish was consigned to the sacrifice zone. Those scattered beyond the Wall were regarded by those inside the Wall, if they were regarded at all, with an uneasy combination of bafflement and pity.

In Plaquemines it was a matter of faith, if not scientific proof, that the Wall was responsible for the devastation wrought by Hurricane Isaac in 2012. While New Orleans experienced minor street flooding, parts of Plaquemines beyond the Wall lay below seventeen feet of water. The corps blamed the disparity on the path taken by the storm, holding the Wall entirely innocent of blame. Plaqueminers believed the Wall trapped the storm surge in their parish as a dammed river will turn a valley into a lake.

"We never had water here before the Wall," said Kermit
Williams Jr., standing on his family land at Wills Point, the
site of the proposed Mid-Breton Sediment Diversion, across
the river and about ten miles south of Warren Lawrence's
dream house. This far down the east bank, the parish ran
a single property wide. Backyards terminated at the parish levee, which defended against the encroaching marshes
of Breton Sound Basin; front yards met the highway that
trimmed the base of the federal levee. Here the Mississippi
was both invisible and oppressive, a tiger in a cloaked cage.
It could not be seen from the ground, even though it was,
on this particular afternoon, seventeen feet higher than the
ground. The crowns of oil tankers, passing like spaceships,
crested above the levee's rim. The saturated flood wall oozed
water that collected in ominous puddles along the highway.

The assemblage of structures on Williams' parcel made a
living tableau of the parish's century. Several hundred yards
from the road lurked the ruined husk of a three-bedroom
house overgrown with tangles of sunbaked vines, cypress
trees, and a vibrating nest of honeybees. It had belonged to
Williams' grandfather; Williams' father was born in its living room in 1910. Before it stood a green stucco house, built
in 1949 on a low foundation of cement blocks, Williams'
home until Hurricane Isaac. Williams lived with his daughter in a third house, closest to the highway. Like most of the
properties along this *Lorax*-like stretch of the parish, it stood
on stilts more than twenty feet high.

Williams commiserated about the brutal history of
heavy-handed interventions in the parish with his neighbors, the brothers Danny and John Hunter. The men agreed
that the parish could survive Nature, but it might not survive

the State of Louisiana. Their litany of grievances began with the bombing of the levee at Caernarvon, a few miles upriver, a shortsighted gambit to spare New Orleans from flooding. The event sparked such fresh daily outrage among Plaquemines residents that a visitor might assume it had happened after Katrina or even Isaac, but this original sin occurred on April 29, 1927. In the 1960s came the construction of the Mississippi River Gulf Outlet, a dagger through the heart of the marsh that contributed to the thirteen feet of water that inundated the Hunters' childhood home after the assault of Hurricane Betsy in 1965 ("It was pretty traumatic," said John, more than fifty years later). The following decades saw the abandonment of a series of smaller diversions on the east bank, each another broken promise from the state. This history made the construction of the Wall feel less like a fresh betrayal than the physical manifestation of a psychic boundary between the saved and the forsaken that had existed for nearly a century.

The Hunters spoke about the master plan in the cadence of undecided voters. They recognized the need to fortify the marshes, given the ever-increasing projections of sea level rise. "If it builds land, I'm for it," said John. "Everyone's for building land," agreed Danny. They could recall a time, not long ago, when they could catch mangrove snapper in New Orleans East, before speckled trout appeared in the shipping canals, when oysters were plentiful on the east bank.

Yet they had internalized the arguments made by the Save Louisiana Coalition, the only nonprofit opposed to the master plan, which represented the movement of Louisiana fishermen desperate to stop the government from trying to save them. Although the fishermen had been making argu-

ments against the diversions for years, the winter of 2019—
the wettest the Mississippi valley had seen in one hundred
and twenty-four years—provided hard evidence for their
apocalyptic predictions. Everything they feared from the diversions had come true, just a few miles upriver.

As the Mississippi has swollen at a terrifying rate, the
corps had opened a different kind of diversion: the Bonnet
Carré Spillway, which functions as a release valve when the
lower Mississippi comes close to overtopping its levees. Built
in response to the 1927 flood, thirty river miles above New
Orleans, the spillway is a seven-thousand-foot stretch of levee composed of seven thousand needles. Each needle is the
trunk of a longleaf pine tree coated in creosote. When the
needles are lifted—manually, one at a time—the spillway
gushes into Lake Pontchartrain. The corps had opened the
spillway a dozen times since 1927, and never in consecutive
years. In 2019 it was opened twice, for a cumulative one hundred and twenty-three days, easily a record. The river water
pulsed past New Orleans to the Gulf Coast, collapsing oyster populations and stranding hundreds of dolphins. The
Department of Commerce declared a federal fisheries disaster. Lawsuits against the corps were filed by the State of
Mississippi, several cities on the Mississippi coast, and environmental groups on behalf of leatherback sea turtles, pallid
sturgeon, and West Indian manatees. "For six years," said
George Ricks, the director of the Save Louisiana Coalition,
"I've been carrying on, to anyone who would listen, that our
industry, our community, our way of life is being threatened.
But nobody paid attention until Flipper started dying." The
toxicity of the Mississippi River frightened the Hunter brothers. They were willing to sacrifice the entire fishery to save

the land, but they couldn't stomach the thought of the river poisoning the marsh.

Most puzzling of all to the Hunters was the state's emphasis on long-term benefits. "The scientists might be right," said John, "but the plan looks ahead fifty years in the future. We don't have fifty years. We need it done now." Even the fifty-year window was misleading, because every five years the clock started anew. The master plan was perpetually, implacably forward-looking. It was a model for the kind of governmental response that galloping climate change demands: an agenda that combined mitigation and adaptation while retaining the flexibility to respond to unforeseen developments, whether positive or catastrophic. It was the rare example of legislation in which the costs were borne immediately and the greatest benefits would not accrue until after the deaths of the elected officials who authorized it. The corps had managed the Mississippi River for nearly a century. The master plan intended to manage the coast for longer.

The Mid-Breton Sediment Diversion would require the states to blast through Danny Hunter's property. The diversion would be excavated adjacent to his property line; the state highway would be rerouted through his backyard. The brothers used the word "reparations." But the value of the land dwarfed its real estate assessment. The rich alluvial soil had given Hunter what he figured was one of America's most fertile backyard gardens, with a profusion of Creole tomatoes, eggplants, snap beans, cucumbers, and squash. Every spring his grove of satsuma and naval-orange trees produced such an abundance of fruit that branches snapped under the weight. In the back of the property, Danny had dug a long pond that he stocked with crawfish, which the raccoons liked to poach. Owls, cardinals, and blue jays nested in the live oaks. The landscape might not have resembled anything that had stood there even thirty years earlier but it masqueraded convincingly as a panorama of lush, pristine wilderness.

When Danny wanted to escape, he strolled along the back levee, fringed with yellow wildflowers, and gazed into the marsh. Mattresses of clover dotted with white buds plunged into a plaid of Roseau cane, yellow palm, and skeletal stands of cypress, killed by saltwater intrusion. The ground looked solid, but in most places it would melt under the pressure of a boot. "Serenity," Kindra Arnesen had said.

"This is my peaceful place," said Danny. "When I'm in the right frame of mind, I get on my knees."

Geographically and ideologically, the Hunters' arpent of paradise stood between Kindra Arnesen and Albertine Kimble, who before her retirement had served as the parish's coastal program manager. Kimble was the only resident of the east bank unprotected by the parish levee, which terminated just south of Danny Hunter's property. The federal levee still protected her from the Mississippi, but even that was too much for her. "It's going to flood," she said, in the darkened living room of her elevated bunker, while the television flashed portentous scenes of the river at flood stage. "I'm not protected by the back levee or the Wall. I'm not protected by the false sense of security that everybody else has. I'm not protected."

In the Kübler-Ross flood cycle Kimble had moved beyond flood denial, flood anger, flood bargaining, flood depression, and flood acceptance to a new stage: floodlust. The diversions, she thought, weren't nearly sufficient. "You'll never be able to imitate what this river does with just two structures. You *can't*. If we want to mimic the river, we need to take the levee out altogether." She spoke with nostalgia of a time before the federal levee was built—a time decades before her birth. She fantasized about living at Fort de la Boulaye, the territory's first European settlement, founded by French explorers in 1699. (The fort, a few miles downriver, flooded constantly and was abandoned after seven years.) She wished that instead of building the levee higher, they lowered it to four feet and just "let 'er rip."

"Wouldn't that be nice? Everything would be naturally moving around." Yes, the river would submerge her property beneath twenty feet of water. "But I'll sacrifice myself to save Louisiana. Man is the greatest enemy. We've tried to control something that Mother Nature did for millennia, building up levees so that man can live here. But the Army Corps is not in control. God is."

Kimble's house, which she had built fourteen feet off the ground, had grown over the years. After Katrina she raised it to seventeen feet; after Isaac, to twenty-two feet, higher than the federal levee. It was still not quite as high as her only neighbor, Ray Givens, a live oak more than half a millennium old that reached its ancient limbs over her like a parent shielding a child. It, too, was slowly dying from saltwater intrusion. Four generations of Kimble's family had lived within a few miles, and none had flood insurance. Kimble did once take out a policy, only to cancel it six months before Katrina on the advice of her insurance agent. He told her, "If you flood, the whole parish will flood." He was close. "The storms," she said, "are winning."

Kimble had received death threats for her advocacy of the diversions but she continued to make trouble. At a party at a local marina celebrating the opening of oyster season, she confronted the leader of the industry association. Immediately she found herself surrounded by "strong, scary" oystermen. She respected them—they worked hard and invested in themselves—but she wanted them to realize that they were fighting against their own self-interest. "Can anybody here tell me that you don't need any fresh water for your oysters?" she asked them. She did not get anywhere with them. "You're not going to change these people's minds," said Kimble. "So they're going to have to adapt. The main issue here is change. Nobody wants change."

The most dramatic change in recent history began during Mardi Gras in 2012 when, just beyond the terminus of the federal levee, the Mississippi burst through the east bank. It created a smallish crevasse at first but after a couple of years the breach widened to forty feet, capturing one percent of the river's volume, and the Coast Guard formally recognized Mardi Gras Pass as a distributary of the Mississippi River. Despite protests from local fishermen, the state declined to plug the crevasse. It had, after all, furthered the master plan's land-building goals; Mardi Gras Pass would alone save the state $1.4 billion. Danny Hunter had firsthand experience of its land-building prowess. Seven years after it had opened, he had cruised an inboard duck boat through a neighboring bayou that he'd traversed dozens of times, where fifty-foot shrimp boats traveled without obstruction, when he ran aground on a newly formed sandbar invisible beneath the water's surface. He was thrown against the bow, breaking his rib.

Mardi Gras Pass had already introduced more than one million cubic yards of sediment into the marsh and, in a lavishly allegorical gesture, had torn through a road used by a Dallas-based oil and gas company—a defendant in the lawsuits filed by the coastal parishes. The pass had also destroyed the oyster habitat on the east bank beyond hope of recovery.

Albertine Kimble had hunted ducks since she was eleven years old. After Mardi Gras Pass broke through, the marsh behind her house clogged with water hyacinths, an invasive, purple-flowering swamp plant that carpets the surface of freshwater canals, repelling ducks. She accepted the trade-off.

"I like seeing all this vegetation," she said. "I like seeing the banks rise. I like seeing the new willow trees. I like seeing the boats running aground. It's awesome."

A few miles upriver, along the southern flank of the Wall, John Lopez, the author of the strategy behind the master plan, and a coastal ecologist named Theryn Henkel were speeding on an airboat down a man-made channel to visit a man-made swamp that was becoming a man-made forest. Airboat travel feels less like boat travel than air travel: you glide over a changeable terrain of open water, swamp, grass, island, rarely suffering so much as a bump or jostle. It gives you a sense of being ruler of all that you see. Lopez wore heavy earmuffs, the kind used by helicopter pilots, to mute the clamor of the dual fan engines. At the airboat's approach a succession of alligators, startled out of their ruminations, flopped into the water like divers in a Busby Berkeley musical.

Lopez was about the only person who could speak with

total candor about southern Louisiana's future. He was no longer employed by the corps and did not dignify political sensitivities. Unlike CPRA's engineers, he was willing to concede that Louisiana's remaining coastal wetlands could not be saved. He accepted the conclusions of a 2020 study led by the Tulane geologist Torbjörn Törnqvist, who found that the present-day rate of relative sea level rise has already surpassed the tipping point at which the drowning of the marsh is unstoppable. (Asked by the New Orleans *Times-Picayune* to translate his findings for the general public, Törnqvist said, "We're screwed.") Törnqvist expected the Gulf of Mexico would ultimately swallow the lower third of the state, turning Baton Rouge into a coastal port and New Orleans, if it exists, into an island city. Still, like Törnqvist, Lopez believed the master plan was worth it, if only to buy time. "Having a few more decades," said Törnqvist, "could mean the difference between managed retreat and complete chaos."

Unlike many of the pro-diversion activists in the state's environmental organizations, Lopez agreed that the fishermen would suffer. Even under the rosiest projections for the diversions, the future coast would be "skeletal" compared with its current form, but "functional," at least in economic and ecological terms. He expected to preserve the state's highway systems, railroads, port and energy facilities, and even its fisheries, albeit in some altered condition. "We can expect that there will be crab, shrimp, and oysters," said Lopez. "We just can't say where or how much." Despite his nostalgia for family fishing trips in Plaquemines Parish, catching speckled trout near Buras, where Kindra Arnesen had dredged for oysters, he did not harbor a tremendous amount of sympathy for the fishermen. "I'm not saying the transition is going to be easy or that it won't cost money. But I'm also not saying that the state or anyone else is responsible for helping." Just because your father was an electrician, said Lopez, and your father's father was an electrician, doesn't mean that *you* need to be an electrician or that the government should incentivize you to be one. The Louisiana fishermen may have been one of the first groups to be passed over by climate policy, but they wouldn't be alone. They would soon be joined by coal miners, offshore roughnecks, long-haul truck drivers, Sonoran farmers, Miami Beach condo owners. Lopez felt for them, he did. But he felt for everyone else more.

The canal followed by Lopez's airboat ran perpendicularly from the Mississippi River. The corps blasted through the levee here in 1991. Formally it was known as the Caernarvon Freshwater Diversion Structure; informally it was three sewer pipes that pierced the federal levee, ran under the highway, and fed into the swamp. Caernarvon had blasted open an earthen and a psychological wound: it was at this very site, in 1927, that Louisiana's power brokers had bombed the levee to spare New Orleans from the great flood, the original sin responsible for nearly a century of trauma and mistrust of government. "Where do they get the authority to drown us out, to deprive us of our homes and our living?" a resident protests, in *Rising Tide.* "We won't stand for it. We should die fighting for our rights." That generation of Plaqueminers demanded reparations too and, despite promises, never received them, an insult that haunted their heirs, no matter their opinion of the master plan—Kindra Arnesen, the Williamses, the Hunters, George Ricks, and Albertine Kimble.

The explosions in 1927 lasted ten days, requiring thirty-

nine tons of dynamite. They opened a flow of a quarter million cubic feet of water per second through the parish—a
Superdome of water every eight minutes and twenty seconds.
The crevasse was not sealed for a year. Despite the ghost of
1927, the 1991 diversion was passionately endorsed by local
oystermen, whose crop had declined for decades because
of saltwater intrusion. When Caernarvon opened, the inundation of river water destroyed oyster production in the
neighboring marsh. But it boosted production across Breton
Sound Basin, as salinity levels returned to historic averages.
Though Caernarvon remained a sore point for local fishermen, Lopez considered it proof of concept: a showcase for the
power of human ingenuity to build an environment that is,
by all appearances, natural.

It was expected that Caernarvon would reduce salinity in
the marsh. The great shock was that it also began to build
land. Unlike the master plan's diversions, Caernarvon was
not meant to capture sediment; for land-building purposes,
it could not have been more poorly designed. It was situated
at a bend in the river where the water flowed quickly, limiting
sediment buildup; it operated intermittently; and it siphoned
from the river's surface, where sediment concentration was
lowest. Yet Caernarvon had nevertheless managed to perform a stunning magic trick.

The airboat pivoted sharply onto a narrow watery trail
through denser forest. Lopez and Henkel had taken to calling
this passage Bayou Bonjour. Among the names erased from
NOAA maps were Bayous Long, Caiman, and Tony; Lopez
and Henkel hoped that Bayou Bonjour would be the first of
many names to be added by the diversions. Bonjour was not
technically a tributary or a creek; it was the final vestige of

a lake remaining between two lobes of land that, like lovers approaching a kiss, had drawn toward each other.

Bayou Bonjour debouched into a steaming marsh indistinguishable from thousands of others in southern Louisiana. This vast terrain, about twice the size of Central Park, was called Big Mar. It was used as a sugar plantation in the early nineteenth century; six shacks housed one hundred slaves. Since the 1927 explosions it had been perpetually inundated, the greatest part of it a brackish lake that grew saltier each year. But it was not a lake any longer. There remained puddles and elongated pools, though none more than three feet in depth. Suspended in the shallow water were aquatic grasses that took the form of green Mardi Gras wigs, tattered velvet, elephantine dill. The roots trapped sediment like weirs. The mud clotted until it surfaced as islets, sprouting exclamatory tufts of giant cut-grass, named for its razor-edged quills, which draw blood. Saplings colonized the accreting land, led by black willow, which shoot up thirty feet within a few years. "I don't think anyone in their wildest dreams imagined that there'd be a forest here," said Lopez. But a forest loomed before him.

It began at the edge of the marsh on land that fifteen years earlier was open water. The Caernarvon diversion had created more than eight hundred solid acres in Big Mar alone. "The whole Big Mar is really restored," said Lopez, gazing upon the scene with pride. The proximity of so many habitats had attracted a garish array of life. That afternoon Henkel and Lopez spotted a blue heron, a white ibis, roseate spoonbills, redwing blackbirds, egrets, and a twitching mullet clutched in an osprey's talons. The alligators were plump and lazy, having gorged themselves on the rich buffet of prey available along the mudflats. The air swarmed with midges, monarch butterflies tested the bull tongue, and iridescent blue dragon-flies browsed the hyacinth.

"When I'm out here," said Dr. Henkel, "I feel like I'm in Jurassic Park."

To secure their territorial gains and hasten the matura-tion of the forest, John Lopez's Pontchartrain Conservancy had planted thirty-six thousand trees, mainly bald cypress, as well as water tupelo, swamp red maple, and green ash. Most of these were planted by migrant workers, a dozen of whom would plant five thousand trees in a week; there were also plantings performed by local volunteers, though they performed a service less efficient than educational. In the ar-eas of marsh not accessible by boat, airplanes flew overhead circuits like crop dusters, only instead of spraying pesticide, they dropped seedlings. The cypress saplings wore plastic collars as a defense from the nutrias, which liked to gnaw them to death. After bursting from the collars, the trees were crowded by an understory of elephant ears, arrowleaf, and black gum. Deeper in the forest skulked muskrat, raccoon, coyotes, deer, boars. Henkel did Dr. Frankenstein: "It's alive!"

The land built by the diversion was an ecological monster—the product of human engineering, compromise, brute force. Like most man-made things it was unruly, even clumsy; its charms, and its dangers, were accidental and unforeseen. It was not, by any conventional definition, *natural*. But it was alive.

9

THE IMMORTAL JELLYFISH

Four thousand years after Utnapishtim told Gilgamesh that the secret to immortality lay in a coral found on the ocean floor, man finally discovered eternal life in 1988. He found it on the ocean floor. The unwitting discoverer was Christian Sommer, a German marine biology student in his early twenties. He was spending the summer in Rapallo, a small city on the Italian Riviera, where exactly one century earlier Friedrich Nietzsche conceived *Thus Spoke Zarathustra*: "Everything goes, everything comes back; eternally rolls the wheel of being. Everything dies, everything blossoms again."

Sommer was conducting research on hydrozoans, small invertebrates that, depending on their stage in the life cycle, resemble either a jellyfish or a soft coral. Every morning Sommer went snorkeling in the turquoise water off the cliffs of Portofino. He scanned the ocean floor for hydrozoans, gathering them with plankton nets. Among the hundreds of organisms he collected was a tiny, relatively obscure species,

Turritopsis dohrnii. Today it is more commonly known as the immortal jellyfish.

Sommer was staying at the home of a fellow graduate student named Giorgio Bavestrello, who traced his interest in the sea to his relationship with his uncle Benito, a local fisherman. As a child, Bavestrello would marvel at the peculiar species that snagged on his uncle's line: luminescent clams, boring sponges, gorgonians, and jellyfish.

The students kept their hydrozoans in petri dishes on a table in the guest bedroom. After several days they noticed that the *Turritopsis* was behaving in a peculiar manner, for which they could hypothesize no earthly explanation. Plainly speaking, it refused to die. In fact, it appeared to age in reverse, growing younger and younger until it reached its earliest stage of development, at which point it began its life cycle anew.

Bavestrello and Sommer were baffled by this development but didn't immediately grasp its significance. (It was nearly a decade before the word "immortal" was first used to describe the species.) Several biologists at the University of Genoa, among them Bavestrello's adviser Ferdinando Boero, continued to study the species, and in 1996 they published a paper called "Reversing the Life Cycle." The scientists described how the *Turritopsis*, at any moment, could transform itself back to a polyp, the organism's earliest stage of life, "thus escaping death and achieving potential immortality." This finding appeared to debunk the most fundamental law of the natural world: you are born, and then you die.

Boero likened the *Turritopsis* to a butterfly that, instead of dying, turns back into a caterpillar. Another metaphor is a chicken that transforms into an egg, which gives birth to another chicken. The anthropomorphic analogy is that of an old man who grows younger and younger until he is again a fetus. *Turritopsis dohrnii* has been referred to as the Benjamin Button jellyfish.

The publication of "Reversing the Life Cycle" barely registered outside the academic world. One might expect that having learned of the existence of immortal life, humankind would dedicate colossal resources to learn how the jellyfish performed its trick. One might expect that biotech firms would vie to copyright its genome; that a global coalition of research scientists would seek to determine the mechanisms by which its cells aged in reverse; that pharmaceutical firms would try to appropriate its lessons for the development of human medicine; that governments would broker international accords to govern the future use of rejuvenating technology. But none of this happened.

Some progress was made, however, in the quarter century that followed Christian Sommer's discovery. We learned, for instance, that the rejuvenation of *Turritopsis dohrnii* is caused by environmental stress or physical assault. We know that during rejuvenation the hydrozoan undergoes cellular transdifferentiation, a process by which one type of cell is converted into another—a skin cell into a nerve cell, for instance. (The same process occurs in human stem cells.) We have discovered that the immortal jellyfish has rapidly spread throughout the world's oceans in what Maria Pia Miglietta, a professor of marine biology at Texas A&M, called "a silent invasion." The jellyfish was found "hitchhiking" on cargo ships that used seawater for ballast. *Turritopsis* has been observed not only in the Mediterranean but also along the coasts of Panama, Spain, Florida, Brazil, and Japan. The jellyfish is able to survive, and proliferate, in every ocean in the world. The prospect of its global conquest summoned images of a distant future in which most other species of life had gone extinct but the ocean consisted overwhelmingly of immortal jellyfish, a great gelatin consciousness everlasting.

Still we have not figured out how the immortal jellyfish ages in reverse. The reasons for our ignorance are all maddeningly unsatisfying. There are, to begin with, very few specialists in the world committed to conducting the necessary experiments. "Finding really good hydroid experts is difficult," said James Carlton, a professor emeritus of marine sciences at Williams College. "You're lucky to have one or two people in a country." This was an example of a phenomenon he called the Small's Rule: small-bodied organisms are poorly studied relative to larger-bodied organisms. There were significantly more crab experts, for instance, than hydroid experts. And more dolphin experts than crab experts. As in conservation, as in literature, as at Pixar—the charismatic species got most of the attention.

The most frustrating explanation for our dearth of knowledge about the immortal jellyfish was of a more technical nature. The species, like the passenger pigeon, was difficult to culture in a laboratory. (Bavestrello and Sommer, in their impromptu guest-bedroom *Turritopsis* lab, had been extraordinarily lucky.) It required close attention and hours of repetitive, tedious labor; even then, it was under only certain favorable conditions, most of which remained obscure to biologists, that a *Turritopsis* would produce offspring. Only one scientist had cultured *Turritopsis* polyps consistently. He worked alone, without major financing or a staff, in a cramped office in Shirahama, a sleepy beach town in Wakayama Prefecture, Japan, four hours south of Kyoto.

Marine biologists tend to be reluctant to make grand claims about *Turritopsis'* promise for humanity. "That's a question for journalists," said Boero (to a journalist). "I prefer to focus on a slightly more rational form of science."

But Dr. Shin Kubota had no such compunction. "*Turritopsis* application for human beings is the most wonderful dream of mankind," he said. "Once we determine how the jellyfish rejuvenates itself, we should achieve very great things. My opinion is that we will evolve and become immortal ourselves."

This seemed worthy of investigation.

Shirahama means "white beach," which is appropriate because one of the city's main attractions is its crescent-shaped white-sand beach. The city devoted considerable resources to maintaining the whiteness of the sand. Every afternoon municipal workers in blue suits and caps walked the length of the beach with picks and plastic bags, collecting litter and seaweed. The city's vigilance derived from its alarm that for several decades its beach had been disappearing. In the 1960s, when Shirahama was connected by rail to Osaka, the city became a popular tourist destination, and blocky white hotel towers rose along the coastal road. The increased development accelerated erosion, and the famous sand began to wash into the sea. Worried that the town of White Beach would lose its white beach, Wakayama Prefecture began in 1989 to import sand from Perth, Australia, forty-seven hundred miles away. Shirahama has since dumped seven hundred and fifty thousand cubic meters of Aussie sand on its beach, preserving its everlasting whiteness—for now.

Shirahama is full of timeless natural wonders that are failing the test of time. Visible off the coast is Engetsu-to, "Full Moon Island," a sublime arched sandstone formation dating to the Cenozoic era that looks like a doughnut dunked halfway into a glass of milk. At dusk tourists gather at a point on the coastal road where, on clear days, the arch frames the setting sun. Arches are temporary geological phenomena; they are created by erosion, which ultimately causes them to collapse. Fearing the death of Engetsu, the prefectural government sought to grant it immortality by reinforcing it with a harness of mortar and grout. A large scaffold was erected behind the arch and from the shore construction workers could be seen, flyspecks against the sparkling sea, paving the rock.

Engetsu is nearly matched in splendor by Sandanbeki, a series of striated cliffs farther down the coast that drop sharply—at about the height and pitch of the Leaning Tower of Pisa—into turbulent surf. Today the cliffs are one of the world's most desirable suicide destinations. A sign on the precipice served as a warning to those contemplating their own mortality: "Wait a minute. A dead flower will never bloom."

Shirahama is best known, however, for its *onsen*, saltwater hot springs believed to increase longevity. There are a number of onsen along the coast—large, well-appointed baths inside resort hotels, smaller tubs that are free to the public, and ancient bathhouses in cramped roadside huts. You can tell from a block away that you are approaching an onsen, because you can smell the sulfur. Each morning,

Shin Kubota, who was nearly seventy, visited a simple onsen popular with the city's oldest citizens that traced its history back fourteen hundred years. There were two baths, one for each gender, partitioned by a rock wall. The waters that fed it were mentioned in poems that appear in the *Collection of Ten Thousand Leaves*, Japan's oldest poetry anthology, compiled in AD 759. "Onsen activates your metabolism and cleans away the dead skin," said Kubota. "It strongly contributes to longevity."

At 8:30 a.m., Kubota drove fifteen minutes up the coast, skirting the white beach, where the land narrowed to a promontory that extended like a pointing, arthritic finger, separating Kanayama Bay from the larger Tanabe Bay. At the end of this promontory stood Kyoto University's Seto Marine Biological Laboratory, a damp, two-story concrete block. Though it had several classrooms, dozens of offices, and long hallways, the building usually had the appearance of being abandoned. The few scientists on staff spent much of their time diving in the bay, collecting samples. Kubota, however, visited his office every single day. If he didn't, his immortal jellyfish would starve.

The world's only captive population of immortal jellyfish lived in petri dishes haphazardly arrayed on several shelves of a small refrigerator in Kubota's office. The refrigerator was set to seventy-seven degrees Fahrenheit, but sometimes Kubota forgot to close the door and the number crept perilously higher; the building lacked central air-conditioning, and this particular week, in mid-July, the halls had the atmosphere of a greenhouse. Kubota's menagerie contained about one hundred specimens, three to a petri dish. "They are very tiny," said Kubota, the proud papa. "Very cute."

It *is* cute, the immortal jellyfish. An adult medusa is about the size of a trimmed pinkie fingernail. It trails scores of hairlike tentacles. Medusas found in cooler waters have a bright scarlet bell, but more commonly the medusa is translucent white, its contours so fine that under a microscope it looks like a line drawing. It spends most of its time floating languidly in the water. It's in no rush. It has all the time in the world.

Turritopsis dohrnii, like most hydrozoans, passes through two main stages of life: polyp and medusa. A polyp resembles a sprig of fennel, with spindly stalks that branch and fork and terminate in buds. When these buds swell, instead of sprouting flowers, they produce medusas. A medusa has a bell-shaped dome and dangling tentacles. Any layperson would identify it as a jellyfish, though it is not the kind you see at the beach. Those belong to a different taxonomic group, Scyphozoa, and tend to spend most of their lives as jellyfish; hydrozoans have briefer medusa phases. An adult medusa produces eggs or sperm, which combine to create larvae that form new polyps. In other hydroid species, the medusa dies after it spawns. A *Turritopsis* medusa, however, sinks to the bottom of the ocean floor, where it folds in on itself, assuming the jellyfish equivalent of the fetal position. The bell reabsorbs the tentacles and degenerates further until it becomes a gelatinous blob. Over the course of several days this blob, technically a cyst, forms an outer shell. Next it shoots out stolons, which resemble roots. The stolons lengthen and become a polyp. The new polyp produces new medusas, and the process begins again.

For about a quarter century Kubota had spent at least three hours a day caring for his brood. It was grueling, te-

dious work. When he arrived at his office, he delicately removed each petri dish from the refrigerator, one at a time,
and changed the water. Then he examined his specimens
under a microscope. He wanted to confirm that the medusas were healthy: that they swam gracefully; that their bells
were unclouded; and that they were digesting their food. He
fed them artemia cysts—dried brine shrimp eggs harvested
from Utah's Great Salt Lake. Though the cysts are tiny, barely
visible to the naked eye, they are often too large for a medusa
to digest. Kubota, squinting through the microscope, sliced
the egg into pieces with two fine-point needles as a father
might cut his toddler's hamburger into bite-sized chunks.
The work caused Kubota to growl and cluck his tongue.

"Eat by yourself!" he yelled at one medusa. "You are not
a baby!" He laughed heartily. It was an infectious, ratchet-
ing laugh—ha, ha, ha, HA!—that made his round face even
rounder, the wrinkles describing circles around his eyes and
mouth.

Once a month, when Kubota was obliged to attend administrative meetings at the university in Kyoto, he returned
the same night, an eight-hour round-trip, to avoid missing
a feeding. When traveling abroad for academic conferences,
Kubota had to carry the medusas with him in a portable
cooler. (He had lectured about *Turritopsis* in Cape Town;
Xiamen, China; Lawrence, Kansas; and Plymouth, England.)
It was a full-time job, caring for the immortal jellyfish, but
they were not the only focus of his research. He was a prolific author of scientific articles, publishing at the rate of one
a week, many based on observations he made on a private
beach fronting the Seto Lab. Every afternoon, after Kubota
finished caring for his jellyfish, he walked the beach with a

notebook, noting every organism that had washed ashore. It was an unforgettable sight, the solitary figure in flip-flops, tramping pigeon-toed along the four-hundred-yard crescent, hunched over, his floppy hair jogging in the breeze, as he scrutinized the quartz and feldspar. He published his data in papers with titles like "Stranding Records of Fishes on Kitahama Beach" and "The First Occurrence of Bythotiara Species in Tanabe Bay." He wrote a jellyfish-of-the-week column in the local newspaper.

To make all of this time for jellyfish, Kubota had been forced to neglect other areas of his life. He never cooked and tended to bring takeout to work. At the lab, he wore T-shirts—bearing images of jellyfish—and sweatpants. He was overdue for a haircut. And his office was a mess. It did not appear to have been organized since he began nurturing his hydrozoans. The door opened just widely enough to admit a man of Kubota's stature. It was blocked from cracking further by a chest-high cabinet, on the surface of which were balanced several hundred objects Kubota had retrieved from beaches—seashells, bird feathers, crab claws, desiccated coral. The desk was hidden beneath a stack of opened books. Fifty toothbrushes were crammed into a pewter cup on the rusting aluminum sink. Most of the framed pictures on the wall depicted jellyfish, including a drawing done in crayon. When Kubota, who has two adult sons, was asked whether one of his children had made it, he laughed, shaking his head.

"I'm not a very good artist," he said, glancing at his desk, where there was a box of crayons.

The bookshelves that lined the walls were jammed to overflowing with textbooks, journals, and science books, as well as a number of titles in English: Frank Herbert's *Dune*,

The Works of Aristotle, The Life and Death of Charles Darwin.
Kubota first read *On the Origin of Species* in high school. It
was one of the formative experiences of his life; before that, he
thought he would grow up to be an archaeologist. He was then
already fascinated with what he called the "mystery of human
life"—where did we come from, and why?—and hoped that he
might discover the answers in the ancient civilizations. But
after reading Darwin, he realized that he would have to look
deeper into the past, beyond the dawn of human existence.

Kubota grew up in Matsuyama, on the southern island of
Shikoku. Though his father was a teacher, Kubota didn't get
excellent marks at his high school, which had been attended
a generation earlier by Kenzaburo Oe. Kubota didn't study;
he read only science fiction. But when he was admitted to
college, his grandfather bought him a biological encyclope-
dia. It sat on one of his office shelves, beside a sepia-toned
portrait of his grandfather. Kubota read every page of the en-
cyclopedia. He was especially impressed by the phylogenetic
tree, the taxonomic diagram that Darwin called the Tree of
Life. Darwin included one of the earliest examples of a Tree
of Life in *On the Origin of Species*; it is the book's only illus-
tration. "The mystery of life is not concealed in the higher
animals," said Kubota. "It is concealed in the root. And at the
root of the Tree of Life is the jellyfish."

When Kubota was a student, the notion that human beings might have anything of value to learn from a jellyfish
would have been considered absurd. Your average cnidarian
does not, after all, appear to have much in common with a
human being. It has no brain, for instance, nor a heart. It
has a single orifice through which its food and waste pass; it
eats, in other words, out of its own anus. As recently as the

mid-1990s, it was believed that the human genome was considerably larger than the genomes of simple animals—that it might include more than one hundred thousand protein-coding genes. Our biological arrogance was deflated by the findings of the Human Genome Project, completed in 2003, which placed the number of protein-coding genes closer to twenty-one thousand, about the same number as that possessed by chickens, roundworms, and fruit flies. Although the size of a genome does not correlate to genetic complexity, a later study found that cnidarians did have a much more complex genome than previously imagined. "There's a shocking amount of genetic similarity between jellyfish and human beings," said Kevin J. Peterson, a molecular paleobiologist who contributed to that study. From a genetic perspective, he said, apart from the fact that we have two genome duplications, "we look like a damn jellyfish."

Peterson suspected this might have implications for medicine, particularly the fields of cancer research and longevity. He studied microRNAs (commonly denoted as miRNAs), tiny strands of genetic material that regulate gene expression. MiRNAs act as an on-off switch for genes. When the switch is off, the cell remains in its primitive, undifferentiated state. When the switch turns on, a cell assumes its mature form: it can become a skin cell, for instance, or a tentacle cell. MiRNAs are the mechanism by which stem cells differentiate. Most cancers, we have recently learned, are marked by alterations in miRNA. Researchers even suspect that alterations in miRNA may be a *cause* of cancer. If you turn a cell's miRNA "off," the cell loses its identity and begins acting chaotically; it becomes, in other words, cancerous.

"Immortality might be much more common than we

think," said Peterson. "There are sponges out there that we know have been there for decades. Sea urchin larvae are able to regenerate and continuously give rise to new adults. This might be a general feature of these animals. They never really die."

Peterson cited the work of Daniel Martínez, a biologist at Pomona College who was one of the world's leading hydroid scholars. He studied the hydra, a species that resembles a polyp but never yields medusas. Its body is almost entirely composed of stem cells that allow it to regenerate continuously. As a PhD candidate, Martínez set out to prove that hydra were mortal. But his research in the following decades convinced him that hydra can survive forever and are "truly immortal." "It's important to keep in mind that we're not dealing with something that's completely different from us," said Martínez. "Genetically hydra are the same as human beings. We're variations of the same theme."

It was this commonality that excited Peterson. "If I studied cancer," said Peterson, "the last thing I would study is cancer. I would not be studying thyroid tumors in mice. I'd be working on hydra." Hydrozoans, he suggested, might have made a devil's bargain. In exchange for simplicity—no head or tail, no vision, eating out of its own anus—they gained immortality.

Yet most hydroid experts found it nearly impossible to secure financing to figure out whether these peculiar, simple species might help us learn how to fight disease, old age, and death. "Who's going to take a chance on a scientist who doesn't work on mammals, let alone a jellyfish?" said Peterson. "The granting agencies are always talking about trying to be imaginative but the pie is only so big."

Kubota's peers spoke cautiously about potential medical applications in *Turritopsis* research. "It is difficult to foresee how much and how fast *Turritopsis dohrnii* can be useful to fight diseases," said Stefano Piraino, a colleague of Ferdinando Boero's. "Increasing human longevity has no meaning. It is ecological nonsense. What we may expect and work on is to improve the quality of life in our final stages."

Kubota disagreed: "The immortal medusa is the most miraculous species in the entire animal kingdom. I believe it will be easy to solve the mystery of immortality and apply ultimate life to human beings."

Kubota was encouraged by the fact that many of the greatest advancements in human medicine came from observations made about animals that, at the time, seemed to have little or no connection to human beings. In eighteenth-century England, dairymaids exposed to cowpox helped establish that the disease inoculated them against smallpox; the bacteriologist Alexander Fleming accidentally discovered penicillin when one of his petri dishes grew a mold; bacteria discovered in hydrothermal vents in the Adriatic Sea in 1986 by divers from the Woods Hole Oceanographic Institution would be used in the tests that diagnose AIDS, SARS, and COVID-19.

So Kubota continued to accumulate data on his favorite simple organism, every day of his life.

There was a second framed picture on Shin Kubota's office shelf, beside the portrait of his grandfather. It was a class photograph of students at Ehime University, in Matsuyama. Though it was taken four decades earlier, the twenty-year-old Kubota was immediately recognizable—the round face, the smiling eyes, the floppy black hair. He sighed when asked about the photograph.

"So young then," he said. "So old now."

The difference was surprisingly subtle. He had gained perhaps a few pounds in the intervening decades, and though his features were not quite as boyish, he retained the exuberant energy of a middle schooler. His hair was full and naturally jet-black.

Yes, he said, but his hair hadn't always been black. Around his fifty-fifth birthday, he experienced what he called a "scare."

It was a stressful time for Kubota. He had separated from his wife, his children had moved out of the house, his eyesight was fading, and he began to lose his hair. It was particularly noticeable around his temples. He blamed his glasses, which he wore on a band around his head. He needed them to write but not for the microscope, and every time he raised or lowered the frames, the band wore away at the hair at his temples. When the hair grew back, it came in white. He felt as if he had aged a lifetime in one year. "It was very astonishing for me," he said. "I realized I had become old."

But he was doing better now, wasn't he?

"Too old," he said, scowling. "I want to be young again. I want to become miracle immortal man."

As if to distract himself from this trajectory of thought, he opened his refrigerator and removed a petri cup. He held it under the light so he could see the ghostly *Turritopsis* suspended within. It was still, waiting.

"Watch," he said. "I will make this medusa rejuvenate."

The most reliable way to make the immortal jellyfish age in reverse is to mutilate it. With two fine metal picks, he began to perforate the medusa's mesoglea, the gelatinous tissue that composes the bell. After Kubota poked it six times, the medusa behaved much like any stabbing victim: it lay on its side and began twitching. Its tentacles stopped undulating and its bell puckered. But Kubota, in what appeared a misdirected act of sadism, didn't stop there. He stabbed it fifty times in all. The medusa had long since stopped moving. It lay limp, crippled, its mesoglea torn, the bell deflated. Kubota looked satisfied.

"You rejuvenate!" he yelled at the jellyfish. Then he started laughing.

Every morning he checked on the stab victim. On the following day, the depleted, gelatinous mess attached itself to the floor of the petri dish; its tentacles bent in on themselves. "It's transdifferentiating," said Kubota. "Dynamic changes are occurring." By the fourth day the tentacles were gone. The organism ceased to resemble a medusa; it looked like an amoeba. Kubota called it a "meatball." By the end of the week, stolons had begun to shoot out of the meatball.

Causing physical distress to induce rejuvenation was, in a sense, cheating; the process occurs naturally when the medusa grows old or sick. In his research Kubota had found that a colony, left to its own devices, will regenerate itself in intervals as brief as one month. He had also learned that certain conditions inhibit rejuvenation: starvation, large bell size, and water colder than seventy-two degrees. In 2019, Maria Pia Miglietta's team at Texas A&M, in partnership with Stefano Piraino in Lecce, used an RNA-sequencing method to identify the biological processes responsible for rejuvenation. It was the first attempt to understand *Turritopsis*' life-cycle reversal at the genetic level; the authors boasted that their findings indicated "possible avenues of future research on developmental strategies, cell transdifferentiation, and aging." Kubota's dream was a step closer to realization.

"Human beings are so intelligent," said Kubota. But he added a caveat. "Before we achieve immortality," he said, "we must evolve first. The heart is not good."

It seemed at first that he was making a biological argument: that we needed to design bionic hearts for longer, artificial lives. But he wasn't speaking literally. By heart, he meant the human spirit: "Before humans can evolve, they must first learn to love nature. When I was a child, I was surrounded by nature. No longer. We live in a mechanical, technical world. People only know about nature through their computers. Today the countryside is obsolete. In Japan, it has disappeared. If this continues, nature will die. There are already signs that it is too late: global warming, overpopulation of the human race, radioactive isotopes found in deep ocean waters. There are unseen chemicals everywhere. We are in the garbage."

Human beings are intelligent enough to achieve biological immortality, he was saying. But we don't deserve it.

"Self-control is very difficult for humans," he continued. "To solve this problem, spiritual change is needed."

This was why, in the years following his "scare," Kubota began a second career. In addition to being a researcher, professor, and guest speaker, he became a songwriter. Kubota's songs were featured on national television, played on karaoke machines across Japan, and made him a minor Japanese celebrity—the Japanese equivalent of Bill Nye the Science Guy.

It helped that in Japan, the nation with the world's oldest population, the immortal jellyfish had an exalted status in popular culture. Its reputation was boosted in 2003 by a television drama, *14 Months*, based on a bestselling novel. Yuko is a thirty-five-year-old newscaster whose career is in decline. She meets a ten-year-old who claims to be the daughter of Yuko's childhood best friend. The girl offers Yuko a potion extracted from the immortal jellyfish. Upon drinking the potion, Yuko begins to age in reverse. Her career revives; her love life improves. Then the twist: she realizes that the ten-year-old is in fact her old friend, aging in reverse, and that the process of rejuvenation will end with Yuko's returning to infancy. Scenes were shot at the Seto Laboratory.

Since *14 Months*, Kubota had appeared regularly on television and radio shows. Highlights from his reel included an episode of *Morning No. 1*, a Japanese morning show, devoted to Shirahama. After a segment on the onsen, the hosts visit Kubota at the Seto Aquarium, where he talks about *Turritopsis*. "I want to become young too!" shrieks one of the hosts. On *Love Laboratory*, a science show, Kubota discusses his experiments while collecting samples on the Shirahama wharf. "I envy the immortal medusa!" gushes the hostess. On *Feeding Our Bodies*, Kubota is filmed staring through his microscope at a medusa. "Among the animals," he says, turning sharply to face the camera, "the immortal jellyfish is the most splendid." There followed an interview with hundred-year-old twins.

No television appearance was complete without a song. For his performances, he transformed himself from Dr. Shin Kubota, erudite marine biologist in jacket and tie, into Mr. Immortal Jellyfish Man. His superhero alter ego had its own costume: white lab jacket, red gloves, red sunglasses, and red rubber hat, designed to resemble a medusa, dangling rubber tentacles. With help from one of his sons, an aspiring musician, Kubota had released six albums. Many of his songs were odes to *Turritopsis*. These included "I Am Scarlet Medusa," "Life Forever," "Scarlet Medusa—an Eternal Witness," "Die-Hard Medusa," and his catchiest number, "Scarlet Medusa Chorus."

> *My name is Scarlet Medusa,*
> *A teeny tiny jellyfish.*
> *But I have a special secret*
> *that no others may possess:*
> *I can—yes, I can!—rejuvenate.*
>
> *When it looks like the end,*
> *I revert to young polyp—1, 2, 3*
> *And restart my life from the beginning.*
> *You all only live once,*
> *so cherish your life for the present.*
> *Oh, hi-ho, scarlet, scarlet.*

Other songs apotheosized different forms of marine life: "We Are the Sponges—a Song of the Porifera"; "Viva! Variety Cnidaria"; and "Poking Diving Horsehair Worm Mambo." There was also a song called "I Am Shin Kubota."

> *My name is Shin Kubota,*
> *Associate professor of Kyoto University.*
> *At Shirahama, Wakayama Prefecture*
> *I live next to an aquarium*

Enjoying marine biology research.
Every day, I walk on the beach
Scooping up with a plankton net,
Searching for wondrous creatures,
Searching for unknown jellyfish.
Dedicate my life to small creatures,
Patrolling the beaches every day,
Hot spring sandals are always on—
Necessary item to get in the sea.
Scarlet medusa rejuvenates.
Scarlet medusa is immortal.

The songs have put him in high demand as a speaker at elementary and high schools. Science classes regularly made field trips to the Seto Lab to meet Mr. Immortal Jellyfish Man. On this particular week he was visited by a group of one hundred eleven-year-olds who had prepared speeches about *Turritopsis*. The group, too large to visit Seto, sat on the floor of a ballroom in a local hotel. After the children made their presentations ("I have jellyfish mania!" one girl exclaimed), Kubota leaped to the stage. He was a natural teacher. He spoke loudly, with great animation, peppering the children with questions. How many species of animals are there on Earth? How many phyla? The karaoke video for "Scarlet Medusa Chorus" was projected on a large screen and the giggling children sang along.

Kubota did not go to such lengths simply for his own amusement, though it was obvious that he enjoyed these performances immensely. Nor did he consider his public outreach secondary to his research. It was the crux of his life's work: "We must love plants—without plants we cannot live.

We must love bacteria—without decomposition our bodies can't go back to the earth. If everyone learns to love living organisms, there will be no crime. No murder. No suicide. Spiritual change is needed. And the most simple way to achieve this is through song. Biology is specialized." He brought his palms within inches of each other. "But songs?"

He spread his hands far apart, as if to indicate the size of the world.

After Kubota finished work, he headed to a karaoke bar. He sang karaoke for at least two hours every evening. He owned a 1611-page book, with dimensions somewhat larger than the yellow pages and even denser type, listing every song available in Japan's karaoke bars. His goal was to perform a number from every page. When he selected a song, he underlined it in the book. A flip through the volume revealed he had already surpassed his goal.

"When I perform karaoke," he said, "another part of the brain is used. It's good to relax, to sing a heartfelt song. It's good to be loud."

His favorite karaoke bar was Kibarashi, which translates loosely to "recreation" but literally means "fresh air." The bar stood at the end of a silent, dark residential street, away from the coastal road and Shirahama's other main commercial stretches, distinguished only by a shoebox-sized sign decorated with an illuminated microphone. Upon opening the door, one encountered what resembled a living room—couches, coffee tables, posters of Japanese singers, pots of plastic flowers, goldfish in small tanks. A low, narrow bar ran along one wall. A karaoke video of a tender Japanese ballad played on two televisions that hung from the ceiling. Kubota stood facing one of them, microphone in hand, swaying side to side, singing full throatedly in his elegant mezzo-baritone. The bartender, a woman in her seventies, sat behind the bar, tapping on her iPhone. Nobody else was there.

Kubota sang for two hours—Elvis Presley, the Beatles, and countless Japanese ballads and children's songs. Upon request, Kubota sang his own songs, seven of which were listed in his karaoke book. Kibarashi's karaoke machine belonged to an international network of such machines. The computer displayed statistics for each song, including how many people in Japan had selected it in the past month. No one had selected Kubota's songs.

"Unfortunately, they are not sung by many people," he said. "They're not popular, because it's very difficult to love nature."

$$\oplus$$

A few days later Kubota developed a bacterial infection in his eye. He couldn't see clearly enough to look through his microscope. When he called to cancel his scheduled meetings, he apologized repeatedly.

"Human beings very weak," he said. "Bacteria very strong. I want to be immortal!" He laughed his hearty laugh.

Turritopsis is also very weak. Despite being immortal, it is easily killed. *Turritopsis* polyps are largely defenseless against their predators, chief among them sea slugs. They are easily suffocated by organic matter. "They're miracles of nature, but they're not complete," said Kubota. "They're still organisms. They're not holy. They're not God."

And their immortality is a question of semantics. "That word 'immortal' is distracting," said James Carlton, the professor emeritus of marine sciences at Williams. "If by 'immortal' you mean passing on your genes, then yes, it's immortal. But those are not the same cells anymore. The cells are immortal, but not necessarily the organism itself." To complete the Benjamin Button analogy, imagine the man, after returning to a fetus, being born again. The cells would be recycled, but the old Benjamin would be gone; in his place would be a different man with a new brain, a new heart, a new body. He would be a clone.

Still Kubota pressed on, in the conviction that the immortal jellyfish would one day restore his youth. He did not fear defeat. He feared easy victory: that the lessons of the jellyfish would be exploited too soon, before humanity was capable of harnessing the science of immortality in an ethical manner. "We're strange animals," he said. "We're so clever and civilized, but our hearts are primitive. If our hearts weren't primitive, there wouldn't be wars. I'm worried that we will apply the science too early, like we did with the atomic bomb."

This line recalled something he had said earlier that week while watching a music video for his song "Living Planet— Connections Between Forest, Sea, and Rural Area." It was shot in Shirahama by his eighty-eight-year-old neighbor, a retired employee of Osaka Gas Company. Kubota's lyrics were superimposed over a sequence of local scenes, in the manner of a generic karaoke video. There was Engetsu, its mossy arch jutting oak and pine trees; cherry blossoms on the Heisogen plateau; the striated cliffs of Sandanbeki; the private beach at the Seto Laboratory; a waterfall; a brook; a pond; and the cliffside forests that abut the city, so dense and black that the trees seemed to secrete darkness.

"Nature is beautiful," said Kubota, with a pained smile. "If human beings disappeared, how peaceful it would be."

10

GREEN RABBIT

In the final hours of the millennium an albino rabbit was born on the grounds of a French agricultural research institute in the town of Jouy-en-Josas, near Versailles. Unlike the bunnies capering about Marie Antoinette's hamlet in the Petit Trianon, those at the Institut National de la Recherche Agronomique were destined to live out their foreshortened existences in a rabbit concentration camp. Down the hill from the laboratories, beyond the abattoir, stood an old, overcrowded hutch. Each of its hundred or so albino rabbits was incarcerated in a wire cage described, by one visitor, as "about two hops long and no hops wide." The millennium bunny was indistinguishable from its peers. Its only discernible character trait was a demeanor so mild that it seemed to have been drugged. The laboratory, unaware the rabbit was destined for global stardom, designated it GFP.014. It was not until four months later, after being visited by a Brazilian artist whose appearance earned him frequent comparisons to Victor Frankenstein, that the rabbit was renamed Alba.

Whether the artist, Eduardo Kac, could be considered Alba's creator would become the subject of a contentious, decades-long, and ultimately inconsequential dispute. The uncontested fact was that the news of his visit to the laboratory, first reported in the local press and later yielding a global bedlam of headlines, polemical essays, and network news features, gave Kac, through a modern form of adverse possession, ownership of the damned creature. Most of the stories led with a photograph that Kac had taken, or doctored, of Alba. Against a white background, the rabbit glowed an alien neon green.

The glow was produced by a gene known as GFP, an acronym for "green fluorescent protein." It had been isolated from the crystal jelly, *Aequorea victoria*, which lived along the Pacific coast. The French scientists had used a synthetic variant that increased the brightness of its glow by two orders of magnitude. They injected the synthetic glow-gene into a fertilized egg cell and implanted it into Alba's mother. (It was through the same process that Revive & Restore hoped to insert passenger pigeon DNA into band-tailed pigeon embryos.) Alba was born with fluorescence encoded into her DNA. This made Alba a chimera (from the Middle English for "wild fantasy"): a jellyrabbit.

Like the crystal jelly, Alba did not glow in daylight or in the dark. Before Kac visited the lab with yellow goggles and a blue light, the French scientists could not be certain that she glowed at all. But Kac confirmed it: under the blue light (which activated the green glow), when seen through the goggles (which blocked the blue light), the rabbit did emit a green aura. The fluorescence expressed itself most vividly in the pinprick intensity of her eyes. Alba's sibling, GFP.015, also glowed, but it protested when Kac tried to hold it. Alba was docile. A photograph was taken of Kac in that moment. He stands before a wall of beige-and-cream swirled tiles out of *Fantastic Planet*, cradling Alba. He clutches her neck perhaps a bit too tightly. Against Kac's gray shirt, buttoned to the collar, Alba's fur radiates bright white, her eye fiery red. Kac, wearing dark glasses, has the forced, closed-mouth smile of a high school principal posing for the yearbook. It would become the portrait by which he would be remembered.

Kac was careful to stipulate, well before Alba became an international art star, that the rabbit herself was not the work of art. The work, which he titled *GFP Bunny*, included not only the creation of the rabbit but her social integration and the public controversy that he expected would ensue. By "social integration," Kac meant that he would adopt Alba as a pet. The adoption would occur in Avignon at an art festival, where Kac would care for Alba in a glassed-in living space for a week. Festivalgoers would be invited to observe the emotional relationship that developed between Kac and his pet. "Alba," he said, "has a cognitive and emotional life that must be acknowledged." Afterward Kac would fly Alba to his home in Chicago. They would join Kac's wife and five-year-old daughter, who would raise Alba "in a loving environment." Photographs and videos of Alba at home with the Kacs would further the humanizing effect. "It's easy to fear what we don't know, that the transgenic is monstrous," Kac told an art critic from *The New York Times*. "But when the transgenic is sitting in your lap, looking into your eyes, the meaning changes."

Kac received an unexpected gift on the eve of the festival when the agronomic institute, INRA, reneged on their agreement, refusing to allow Alba out of her wire prison. This enabled the festival director to vent his "violent displeasure" at this "unjustifiable" act of censorship, and it gave Kac an exquisite demonstration of one of his central justifications for the work: that man-made creatures were not granted the dignity accorded to other living creatures. INRA's decision did more to make Kac's point than a week spent rabbitsitting in a faux Avignon living room ever could have.

"It is a normal organism," Kac told *La Provence*. "It is not a monster!"

INRA would blame the sudden reversal on miscommunication, claiming that its director would never have allowed lab property to go free into the world. The far more likely explanation was miscalculation. The French government scientists had not entered into partnership with Kac out of a spirit of generosity or high-minded artistic impulse. They expected *GFP Bunny* to create a public sensation. They had just believed it would be a positive one.

They had needed good press. The INRA lab at Jouy-en-Josas was a leading research institution in the study of genetically modified organisms, a field under tremendous pressure in France. A series of scandals involving government scientists—most flagrantly the revelation that doctors had knowingly injected hemophiliacs with HIV-tainted blood in the mid-1980s—together with a rising popular movement to protect peasant farmers against the depredations of agricultural multinationals, had turned popular sentiment violently against the promises of novel biotechnologies. Kac's partner at INRA, the director of research, Louis-Marie Houdebine, was an enthusiastic advocate for genetically modified organisms, having published a polemic on the subject for the general public. The insertion of DNA into a foreign species allowed researchers to determine more precisely the function of specific genes. Experimentation on transgenic animals, Houdebine was at pains to make his countrymen understand, could lead to breakthroughs in the treatment of cystic fibrosis, measles, hepatitis C, HIV, Alzheimer's disease, dementia, breast cancer. Houdebine's own research applied genetic engineering to develop a new field of pharmaceutical drugs—vaccines, hormones, antibodies—derived from animal proteins. Houdebine studied the production of novel proteins in the milk of transgenic rabbits. INRA had been producing GFP rabbits for more than a year before it heard from Kac.

Originally Houdebine had agreed to participate in a debate at the Avignon festival on the aesthetic beauty of GFP rabbits. He would use the opportunity to promote the value of genetic engineering. But while Houdebine might have had a feeling for beauty, he had none for art. As Kac could have explained to him, a work of art demanded more than aesthetic harmony; it asked difficult questions to which there were no satisfying, or even knowable, answers.

It quickly became clear that the scientist and the artist had opposing goals. Houdebine wanted propaganda. Kac sought public outrage, the messier the better. The success of *GFP Bunny* hung on the contrast between the cuteness of the bunny and the artificiality of its appearance. The higher the contrast, the greater the horror. As soon as it became clear that Alba would generate more horror than sympathy, INRA realized its blunder and tried to reverse engineer itself out of the project.

Kac had tried for years to make art that troubled the ethical morass of genetic engineering, a subject that had gained greater salience as the Human Genome Project neared completion (the first draft of the human genome was published in February 2001). In *Time Capsule* (1997), a performance conducted in a mock hospital bed in a São Paulo art gallery, Kac inserted a microchip beneath the skin of his ankle—a digital ankle bracelet that tracked his location online. Use of these microchips had until then been reserved for lost pets; the shackle evoked slavery. "Not even the most personal of all biological traits," wrote Kac, "is immune to greed and to technology's omnipresence." In *A-Positive*, held in Chicago the same year, Kac hooked himself intravenously to a robot that extracted oxygen from his blood to ignite a flame; he hoped it demonstrated that "not even DNA or blood are immune to the invasion of the body by technology." In *Genesis* (1999), Kac translated a scrap of Genesis 1:28 ("Let man have dominion over the fish of the sea, and over the fowl of the air, and over every living thing that moves upon the earth") into DNA base pairs. He inscribed this "Genesis gene" in bacteria, displayed in a petri dish at an art gallery in Linz, Austria. Visitors flashed the bacteria with ultraviolet light, introducing mutations, which were translated back into English.

None of these precursors approached the elegance of a green rabbit. No artistic statement was required to understand Alba (though a lengthy one Kac did write). The living room performance would have been superfluous. Even the titular invocation of "GFP" was superfluous. The colliding ironies, the moral tensions, the specter of civilization's in-

creasingly godlike manipulation of organic life—all were tidily contained by the image of the preternaturally glowing
rabbit. Alba was Kac's *Crying Girl*, his *Identical Twins*, his
Campbell's Soup Cans. He replicated the image in silk screens,
paintings, photographs, drawings, poster boards, digital art
projects, and on a flag he raised in front of his house.

The public attacks were as predictable as the glee of head-
line writers: DOES BUNNY ART GLOW TOO FAR?; ENERGIZER
BUNNY; CROSS HARE; FRENCH SCIENTISTS HOPPING MAD
OVER GM RABBIT; ONE SMALL HOP FOR ALBA, ONE LARGE
HOP FOR MANKIND; and the pensive IF ALBA IS A MONSTER,
SO ARE WE ALL. The accusations of animal cruelty were the
easiest to deflect. Kac was responsible for "Alba," but not for
GFP.014; the doe belonged to a global industry of animals
bred, and slaughtered, for scientific research. Kac even won
the support of PETA, which praised him for bringing attention to the plight of lab animals. Still Kac couldn't resist
pointing out that the history of art was soaked in leporine
blood; for centuries painters had sealed their canvases with
rabbit skin glue. "Behind every da Vinci, Velázquez, Goya, or
Picasso," said Kac, "there are countless dead rabbits."

Kac was bored by pat vilifications of the emergent technology. How far, the editorialists asked, should we take our
power to mold other creatures? How extravagantly should
we transform ourselves? The answer, as the history of the
human race has amply demonstrated, was self-evident. We
would go exactly as far as the technology allowed, and strain
to go further. The challenge was to encourage responsible use
(as much as possible), and to avoid grotesque abuses (as much
as possible). The ethical quandaries raised by genetic engineering demanded a sweeping cultural debate that took into

account historical antecedents, the state of the science, and the full range of potential applications. "Since the domain of art is symbolic," wrote Kac, it can "reveal the cultural implications of the revolution underway." Each person would reach his or her own conclusions, but artists should ask the questions.

What did become clear, in the months of scandal and scare journalism that followed Alba's absence from Avignon, was that neither INRA nor Kac knew much about the actual creature that had inspired this public reckoning. "'Alba' doesn't exist," said Louis-Marie Houdebine. "For me, it's rabbit number 5256 or so." Alba could not be said to have existed in any more tangible way for Kac, who did not see the rabbit after their original brief encounter. Kac liked to say that meeting Alba had awoken in him "a deep sense of responsibility," which was a sentiment he could just as easily have expressed before the encounter, or without having met the bunny at all.

At the end of the year, after the original burst of public outrage had ebbed, Kac returned to France to emancipate Alba—or at least to generate a new round of publicity. He did not bother to contact INRA, as he admitted to Christopher Dickey, a reporter from *Wired* who accompanied him on the trip. Instead, Kac spent his time giving interviews, wheatpasting posters of his portrait with Alba all over Paris, and hosting public debates. Audiences divided between those who accused him of being a modern Dr. Mengele and those who confided, "You know, I would like to glow."

While Kac made his rounds, Dickey visited INRA, where he met Alba. He found the doe catatonic in her shit-stained cell.

The most strenuous attacks on Kac tended to come from art critics, journalists, and ethicists who did not realize that Alba, far from being a visitation from an uncanny future, was a creature of the present. As early as 1991, Houdebine had experimented with injecting *human* DNA into rabbit embryos to produce a novel source of protein for medical applications. Use of GFP went back to 1962, when it was first isolated by a team led by Osamu Shimomura, one of three scientists who would win a Nobel Prize for work on the protein. It was first inserted into another species, the ringworm, in 1988. By the time Alba was born, GFP had been used as a biomarker—to trace, for instance, the growth of tumors—in flies, mice, frogs, maize, and slime mold.

The history of transgenic animals went back to 1980, when a herpes virus was successfully injected into mice embryos. The same year, the U.S. Patent and Trademark Office granted its first transgenic animal patents. It has since issued patents for hundreds of species, among them glowing pigs, cats, dogs, sheep, monkeys, and pet fish. The first transgenic rabbit was produced in 1985. When scientists were asked for their view of *GFP Bunny*, their main criticism was that it was "frivolous," as one Harvard biologist put it, because there were "many more important things" that could be done with the fluorescent gene. Alba was old hat.

The phrase "genetically engineered rabbit" approached pleonasm. Pet rabbits are themselves human creations, domesticated through centuries of consumption, transportation, coevolution, and breeding. White New Zealand rabbits, favored by researchers for their temper and reliable health, were the accidental creation of a California breeder named William Preshaw in 1917. As the evolutionary biologist Marc Hauser said of Alba, "This rabbit is not any sillier than a Chihuahua." As with most engineering advances, what had changed was not the act of intervention but the precision and scale with which it was executed.

In the summer of 2002, INRA announced Alba's death. No cause was given. "So rabbits die often," said Houdebine. "It was about four years old, which is a normal life span in our facilities." But Alba wasn't four years old; she would have been two and a half. And New Zealand Whites lived between five and eight years. INRA's rabbits weren't bred to live, however. They were bred to be slaughtered, their organs harvested for research. Kac was skeptical of the pronouncement, suspecting that Houdebine had declared Alba dead to be rid of the constant press inquiries. It seemed likelier, if more prosaic, that like her hundred hutchmates, she had been killed for science.

In the end the circumstances of Alba's death, like those of her birth, were of no consequence. The Alba flag continued to fly from Kac's house; because it served, in the artist's words, as "a beacon of her absence," there was no reason to lower it to half-mast. The *GFP Bunny* project advanced unabated, and Alba's artistic offspring, the prints and shirts and staged debates, proliferated.

Even in this way Alba's fate emulated scientific progress. Shortly before Alba's death, researchers discovered that transgenic GFP animals, after death, continue to glow.

Following *Genesis* and *GFP Bunny*, Kac completed his *Creation Trilogy* with *The Eighth Day* (2001), a domed ecosystem in which every living organism—mice, tobacco plants, zebrafish, amoebas—glowed. For *Natural History of the Enigma* (2009), he created "Edunia," a transgenic pink petunia that expressed Kac's DNA in its petals' red veins. *Cypher* (2009) was a DIY kit that allowed viewers to perform an act of transgenesis themselves. Meanwhile "bio art," a term Kac had coined in an early manifesto, grew into an expansive discipline. It spurred the creation of new art institutes, a steady pace of exhibitions, and a growing volume of critical literature. "It makes no sense to paint as we painted in the caves," said Kac. "It is a new era, and we need a new kind of art."

It was possible to define "bio art" broadly enough to take into account the cave drawings at Lascaux, Leonardo's anatomical sketches, Ernst Haeckel's illustrations of flora and fauna, the land sculptures of Andy Goldsworthy, even the unintentional aesthetic marvels of medical X-rays. Kac defined the term more narrowly, as art that "manipulates, modifies or creates life and living processes," requiring "direct biological intervention." The emerging canon includes the work of Suzanne Anker, founder of the Bio Art Lab at the School of Visual Arts in New York, who creates psychedelic petri dish sculptures of organic material; *Nature?* (1999) by the Portuguese artist Marta de Menezes, who deformed butterflies in the cocoon to give their wings asymmetrical patterns; and *Light, Only Light* (2003) by the Japanese artist

Jun Takita, who planted a three-dimensional reproduction
of his cerebrum with genetically engineered bioluminescent
moss, rendering his brain a glowing garden. Oron Catts and
Ionat Zurr, a husband and wife in Perth, specialize in making "semi-living sculptures," using living tissues to make a
leather jacket (*Victimless Leather*), wings cultured from por-
cine bone marrow (*Pig Wings*), and, in *The Semi-living Steak*
(2000), the earliest example of lab-grown meat, composed
of cells harvested from an unborn sheep. Catts and Zurr
worked with the Cypriot-Australian artist Stelarc to grow a
miniature replica of his ear in 2003; Stelarc has since undergone multiple surgeries in an effort to grow a human ear on
his forearm. The idea was taken from an experiment by the
Harvard Medical School researcher Charles Vacanti, who in
1996 implanted a replica of a life-sized human ear, grown
from cow cartilage cells, onto the back of a lab mouse.

"The function of art is to expose areas of life that we
don't have the proper language to describe," Catts has said.
Bio art, he added, seeks out "zones of discomfort." The bio
artists have little trouble entering a zone of discomfort. It is
more difficult to entice a viewer to remain in it, scrutiniz-
ing its wonders as well as its horrors, and not turn away in
disgust. It is most difficult of all, however, to make a viewer
understand that all of humanity has entered a zone of discomfort, a thoroughly unnatural realm of man-made beasts and
genetic overlords, that we will not exit until we've decorated
it to look like home—like the ersatz living room Eduardo
Kac designed for Alba.

In the years following the Alba controversy, Louis-Marie
Houdebine gave media training lectures to scientists to help
them forestall similar distractions. He hoped to teach other

scientists how to make their work more palatable to journalists, so as to avoid scaring the public. "A consensus on a problem is reached," Houdebine told them, "when society [stops] talking about it."

This was only half-true. When a new problem emerges, experts in the field develop their own consensus. It takes much longer for the implications of major technological change to be grasped by nonexperts. By the time headlines begin to appear, use of the technology has already been widely adopted, its standards of use entrenched. Public outrage was not a response to a new technology. It was an aftershock.

Houdebine and his colleagues had made their own peace with the exploitation of transgenic animals decades earlier, but the rest of society had not begun to grapple with the technology's implications when Kac traveled to Jouy-en-Josas. Kac called his effort to adopt the rabbit "Free Alba," but she had already been freed, conceptually if not physically, and it was that freedom, and the attention it received, that Houdebine resented. Society still has not stopped talking about the ethics of our transgenic future. The conversation has barely even started.

Bio artists borrowed one of science fiction's oldest tricks: they extrapolated the present state of science into a speculative future that has already adjusted to it. Imagine, they asked, what it will be like when chimerical, glow-in-the-dark pets are commonplace; when bodies can be cultivated like gardens; when pigs have wings. But their art, like science fiction, was not prophecy; it could not be. The artists could only respond to a reality that had already arrived. Their creations lagged the science and could never hope to catch up.

This lag doesn't reflect an artistic failure. The lag gives art its highest meaning. It is in the awkward, painful period, between the emergence of a new world and our realization that we already inhabit it, that imaginative art is most desperately needed. Enlightenment lies not in renouncing reality but in seeing it more clearly. Art, even flawed art, helps us to understand our own place in an unfamiliar landscape. It gives language to our most inchoate terrors and desires. It shows us how the soul is worked on by an age of radical upheaval, and how the soul must respond.

ACKNOWLEDGMENTS

The following people deserve special thanks for their help with the foregoing:

Meredith Angelson, James Burnett, Elyse Cheney, Wes Enzinna, Sarah Fineman, Claire Gillespie, Claire Gutierrez, Jon Kelly, Jason Lalljee, Hugo Lindgren, Sean McDonald, Claire Potter, Simon Rich, Theodore Ross, Robert Silvers, Jake Silverstein, Bill Wasik, Dylan Wells, and Sean Woods.

Nathaniel Rich is the author of *Losing Earth: A Recent History*, a finalist for the PEN/E. O. Wilson Literary Science Writing Award, and a winner of awards from the Society of Environmental Journalists and the American Institute of Physics. He is also the author of the novels *King Zeno*, *Odds Against Tomorrow*, and *The Mayor's Tongue*. He is a writer-at-large at *The New York Times Magazine* and a regular contributor to *The Atlantic*, *Harper's Magazine*, and *The New York Review of Books*. He lives in New Orleans.